CONSTRUCTION MATTERS

CONSTRUCTION MATTERS

Georg Windeck

co-editors Lisa Larson-Walker, Sean Gaffney, Will Shapiro

design Leah Beeferman

pH **powerHouse Books** Brooklyn, NY

Construction Matters

Published in the United States by powerHouse Books,
a division of powerHouse Cultural Entertainment, Inc.
37 Main Street, Brooklyn, NY 11201-1021
telephone 212.604.9074, fax 212.366.5247
e-mail: info@powerHouseBooks.com
website: www.powerHouseBooks.com

First edition, 2016

Library of Congress Control Number: 2015951916

ISBN 978-1-57687-778-4

Printed and bound in China by
Toppan Leefung Printing Limited

Design by Leah Beeferman

10 9 8 7 6 5 4 3 2 1

Front and back cover   New National Gallery in Berlin
by Ludwig Mies van der Rohe, under construction.
Photo by architect Rolf D. Weisse, Berlin.
Title page   Farnsworth House in Plano by Ludwig
Mies van der Rohe. Photo by Construction Matters.

für meinen lieben Vater

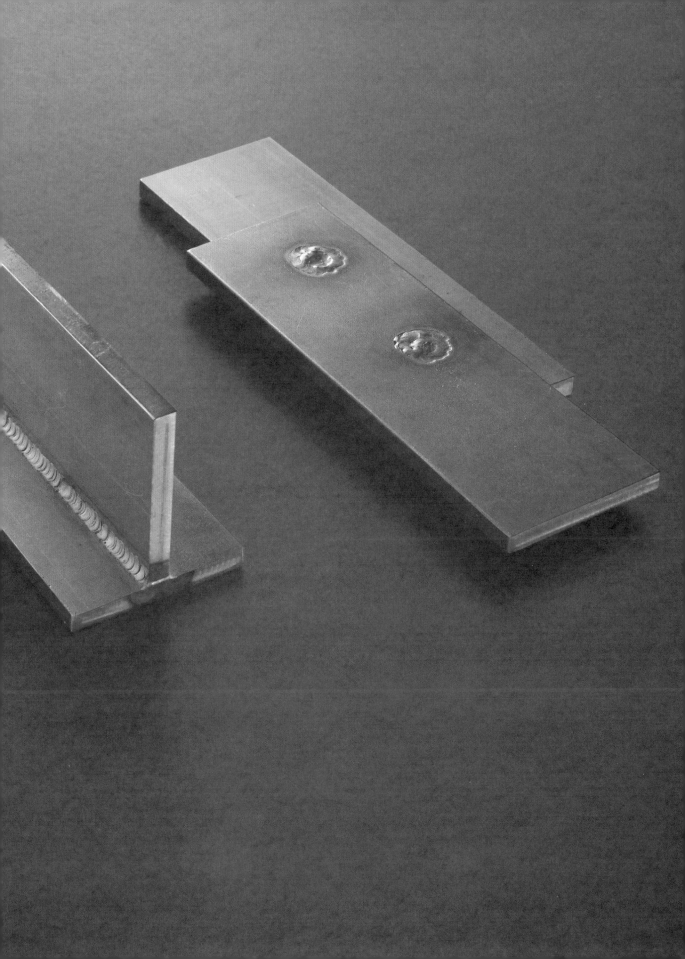

INTRODUCTION

The materials of construction establish the essential shape and character of architectural space. Their structural capacities and aesthetic gesture are inextricably bound to the form of a building. Construction materials are the artistic media of the architect, regardless of how functional, social or site-specific criteria may influence their thinking.

The formal potential of a building material is both generated and limited by its inherent qualities and methods of use. Masonry, concrete, steel, wood: each anticipate a particular structural logic and spatial character. A building is rarely constructed from a single material; yet one particular building method often establishes the main principles according to which an architectural design is generated, determining its fundamental tectonic gesture.

Masonry structures consist of units that are cut from rock or molded from clay; these units are stacked on top of each other to create a wall or pier. Their modular shape determines the proportions and dimensions of the building elements they constitute. The pattern of their assembly establishes the surface texture of the building, in which the construction procedure remains tangible as a permanent architectural feature. The surface of a brick wall creates a comforting appearance because the size of its units relates to the size of a human: The brick dimensions are determined

by the weight a mason can easily hold in one hand; we can imagine ourselves as the builders of such a structure. The surface of a wall erected from large stone ashlars does the opposite. It creates a monumental gesture that radiates the impressive force that must have been necessary to hoist the units into place.

Just as masonry anticipates an architecture of vertical planes, reinforced concrete has the capacity to create horizontal surfaces. It can span and cantilever because of embedded steel reinforcing. The embedded reinforcing bars are as essential for the structural performance of a reinforced concrete member as the concrete itself. But they become an invisible force once the cement and water mixture has been poured over them. The monolithic appearance of the finished member emphasizes its volumetric character and celebrates the plasticity of the material. It can assume virtually any shape. The free articulation of form is limited only by the formwork into which the liquid concrete is poured during construction.

0.4 Concrete is a plastic volume that can assume any shape. Concrete can span and cantilever, emphasizing horizontality as this parking garage demonstrates.
0.5 Masonry is composed of units that are stacked on top of each other. Masonry construction emphasizes verticality and massiveness, as demonstrated by the Monadnock Block in Chicago by Burnham and Root from 1893.

Construction Matters

The shape of a reinforced concrete member can express the flow of forces running through it. Complex curvature may be determined by structural performance and the desire to minimize the required amount of material. Alternatively, the shape of a member can hide or even contradict the flow of forces. A simple rectangular configuration may function structurally just as well as an optimized minimal surface, if a sufficient amount of reinforcing is embedded in strategic locations. Such structures express the capacity of reinforced concrete in a different way: The form contradicts a natural sense of gravity and the force that the composite material resists can be visualized.

The homogenous surface character of concrete supports its volumetric expression. Concrete construction creates a scale-less appearance that does not memorize the construction process like a masonry wall. Only traces of the wooden or synthetic formwork into which the concrete was poured remain as imprints on the completed surface of the building. The finished form conveys the tactile qualities of planks or panels, without allowing them to become a permanent part of the architecture.

In steel construction, the material itself is produced as a linear member, ready to assume a structural function as column or beam. The shape of the member anticipates its load bearing function in a structural system. And the jointing techniques used to connect members define the way in which forces are translated. Joints can make a steel framework rigid or flexible, while expressing the structural characteristics of the assembly as a whole through the language of the detail.

Masonry walls create a vertical border between one side and another; concrete slabs create horizontal boundaries between above and below. A steel framework draws lines into space that crop the surrounding environment, framing the features of the landscape like a painting. A steel structure does not establish any surfaces; it must rely on additional materials to create spatial separation.

Wood construction also composes frameworks from linear parts. It recalls the primal concept of architectural tectonics: A specific hierarchy of columns and beams that establish a built form. Joints between members articulate the structural function of each member within the overall assembly. Wood has a much smaller load bearing capacity than steel. A greater number of members is required to create a sturdy structure. Wooden members are soft: they cannot be connected as rigidly as steel. This makes additional timbers necessary to create a stiff system. The result is a complex arrangement of posts, studs, rafters and bracings that fills space with a dense three-dimensional pattern. The softness that limits the structural capacity of wood also allows it to be tooled easily during assembly on site.

0.6 In steel construction, linear members are jointed to form a structural system. A framework draws lines into space without establishing any boundaries as can be seen in the Shibaura building in Tokyo by SANAA from 2011.

The current use of masonry, concrete, steel and wood was developed over the course of the twentieth century. Many technologies that form the core of the contemporary architect's repertoire were major transformations over the past hundred years. The industrial revolution caused a sea change within the building arts, initiating a new drive within the construction industry towards continual technological progress. Artistic experiments with new materials, jointing techniques, fabrication procedures and structural calculation methods gave rise to the aesthetics of Modernism.

The following chapters present selected innovations that brought about the way we build today. The architectural exploration of these inventions is discussed with case study buildings that are chosen based on an obvious formal relationship between their tectonics and the new technology they incorporate. The case studies constitute a broad spectrum of *contrasting forms of expression*, rather than a balanced variety of architects, times and places. These forms range from massive rectilinear masonry volumes to fragile curved concrete shells, from lofty steel frames to enclosed wooden boxes, spanning prewar architectures in the United States and Europe to contemporary projects in Asia. All these forms are radically conceived from within a material technology. They encourage today's architects to investigate and reevaluate each building method in light of constant transformation.

Construction Matters is not an encyclopedic overview of available technologies and potential applications.

Nor is it a history book, seeking to establish a particular ideology about each construction material based on a pantheon of unsurpassable masterpieces. Rather, it proposes a way of thinking about architecture in relation to technology that transcends a particular building method or invention. *Construction Matters* refocuses the contemporary discourse about architectural aesthetics on the craft of building as a creative endeavor, positioning architecture as a dialogue between artistic concepts and engineering methods.

0.7 A great number of wooden members are connected in a dense arrangement to create a timber structure like the Gamble house in Pasadena by Greene and Greene from 1908.

BRICK MASONRY

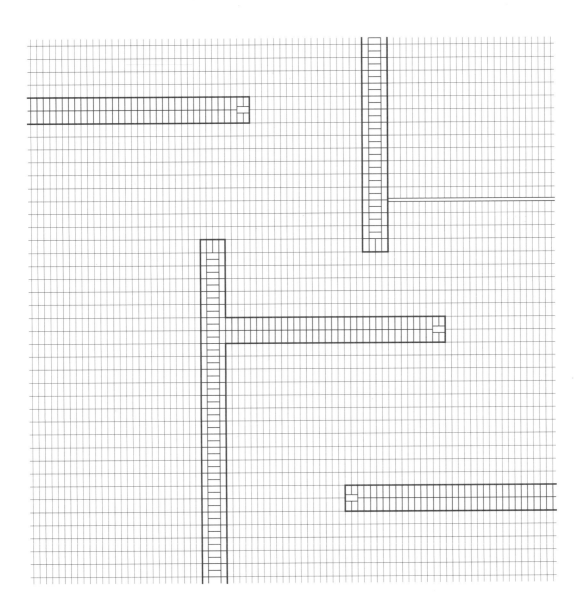

BRICK COUNTRY HOUSE

HOUSE WITH THREE COURTYARDS

MONUMENT TO ROSA LUXEMBURG
AND KARL LIEBKNECHT

HOUSE IN MUURATSALO

BRICK MASONRY

It must have been one of the primal inventions of architecture to take a handful of earth and shape it into a unit that can be reproduced to build a house. Brick masonry is one of the oldest methods of construction. Its basic principle of parts creating a whole is integral to the history of building activity and the definition of architectural space as we know it. From the simple adobe brick types of Ancient houses to the complexity and finesse of Gothic brick structures, the material experienced many technological advancements that evolved over the course of its long history. Techniques such as glazing and burning were invented to increase its durability and strength. New structural systems such as arches and vaults allowed for radically new spatial expression. In light of this history of reinvention and innovation, brick masonry is an ideal construction method to begin a discussion about the formal impact of technological progress on Modern architecture. This discussion must begin with the following questions: What were the technological novelties in the brick masonry of the twentieth century? And how have they influenced the form and expression of brick buildings?

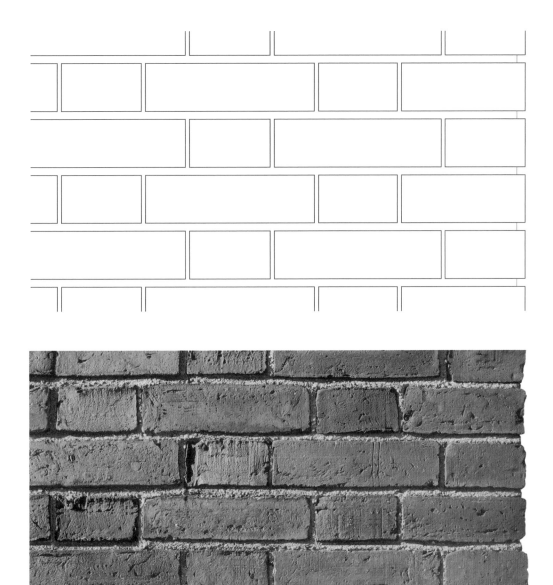

1.1, 1.2 A brick wall displays its basic construction
principle of modular parts creating a whole.

Brick Masonry

EXTRUDED WIRE-CUT BRICK

The production of bricks has been remarkably consistent since they were first used around 5,000 BC. Until the nineteenth century, bricks were made in essentially the same way: a clay mixture was pressed into a form the size of the desired brick, and then air dried or burned. While this method of producing brick was simple, cheap and reliable, it had certain drawbacks. The clay had to be moist in order to be soft enough for its manual placement into the form. As the wet clay dried, it inevitably shrank and became distorted. These irregularities limited the strength of the material, as the stress was distributed unevenly over the brick's surface. Furthermore, the use of individual forms for each brick was unavoidably inefficient with regards to time and storage; with the increase in demand during the Industrial Revolution, such inefficiency needed to be rapidly overcome.

The invention of extruded wire cut brick in the late nineteenth century fundamentally transformed masonry. Higher quality brick could be produced in vast quantity, allowing the construction booms that took place in many cities during the Industrial Revolution. In the new method a very stiff, i.e. dry, clay mixture is forced through a die and cut down to size with a wire into multiple brick sections, instead of using individual molds. The dry mixture minimizes shrinkage during the burning process, producing a very precise and consistent brick. This procedure results in a stronger brick that can bear more unit-stress per square inch. Walls could be built with thinner dimensions due to the increased strength of each individual brick. Where mortar seams had been thickened to compensate for irregularities in the past, they could now be reduced due to the consistent brick dimensions. The masonry thus becomes tighter, sharper, and more compact.

1.3 In the production of extruded wire cut brick,
a very stiff, i.e. dry, clay mixture is forced through
a die and cut down to size with a wire into multiple
brick sections.

Extruded Wire-Cut Brick

BRICK COUNTRY HOUSE

The Brick Country House by Ludwig Mies van der Rohe is one of the most influential brick masonry projects of the twentieth century, despite the fact that it was never built. It embraces aspects of traditional brick masonry in the detailing of how bricks meet and align. Yet the architectonic arrangement of its walls is radically abstract and modern, liberating brick from the formal constraints of its history.

The design concept for the house can be read as a celebration of the formal possibilities inherent in extruded wire cut brick. The sharp lines of the drawings, which allow the walls to dematerialize into an abstract assembly of elements, rely upon the sharp edges of the newly developed brick production method. The degree of detail with which the construction is laid out in the drawings necessitates the consistent dimensions of this industrial product. All together the precision afforded by extruding and wire cutting allows a brick to be regarded as an abstract rectangular prism rather than a familiar building block. So too does the Brick Country House do away with any pictorial notion of what a "brick country house" should be, opening up a poetics that transcends style and historicity.

BRICK STILL LIFE

The perspective of the Brick Country House is a still life composed of different bricks. Like in any still life, the surface on which the objects rest is essential. A painter typically arranges a composition of objects upon a table. The common surface that they share is necessary to define their relation to each other and emphasizes their figurative qualities. Whereas a Renaissance *vanitas* may contain flowers, fruit, and dead animals, Mies' still life is about bricks of various proportions. Instead of using a table, Mies places his volumes on a continuous plinth that is articulated by a sharp edge at its perimeter. This plane literally *elevates* the bricks to an artistic subject. At the same time, it *underlines* the object-like character of the individual parts of the house and emphasizes their specific proportions. The whole building becomes a scale-less cubist expression. Just like the perspective is a still life of volumes that depicts the house in the manner of representational painting, the plan is a composition of lines that resembles an abstract painting. While the perspective shows the walls as planes that suggest solid volumes, they become lines that imply a configuration of spaces in the plan. This drawing states the design concept of the house: an architecture that is made out of only walls.

Mies' approach to brick masonry is clear and direct: Brick is a material from which walls are built. A house that is entirely built from bricks has to consist of different variations of walls.

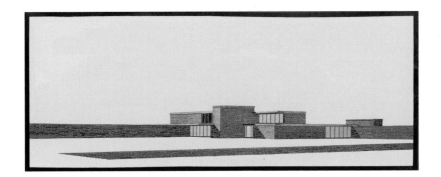

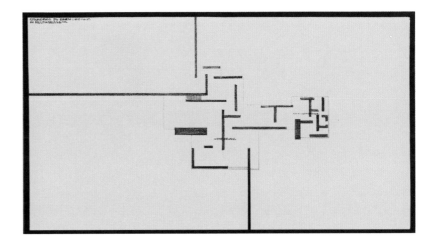

1.4 The perspective drawing of the Brick Country House from 1926 looks like a still life composed of different brick sizes and proportions.
1.5 The plan of the Brick Country House resembles a cubist painting. The arrangement of walls creates an open and dynamic space, while maintaining the traditional logic of brick construction.

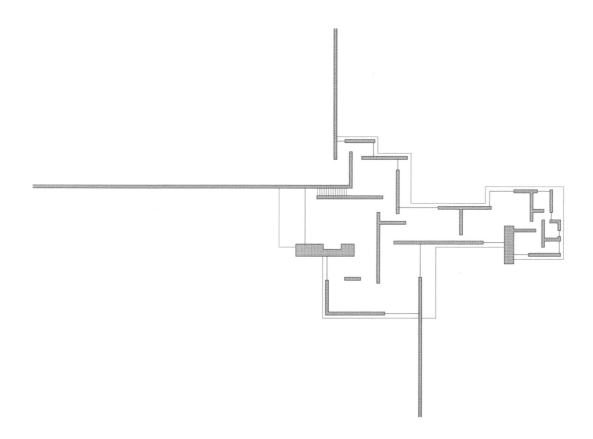

ARCHITECTURE OF WALLS

A perimeter wall with a square or rectangular plan would be a good example for this masonry concept in its most archetypal form: Four walls are necessary to create a load bearing structure that is also an inhabitable enclosure. The configuration of walls connecting in right angles at the corners of a building assures the stability of the structure, resulting in a closed shape.

In the plan of the Brick Country House the walls are free standing and project through space independently. The drawing reveals a lightness and dynamic gesture that is altogether new for masonry in its traditional sense. With extruded and wire cut brick, a wall no longer needs to be a huge mass; because of the hardness of the material the structural width of the wall can be reduced and it becomes a thin linear element. The architectural purpose of Mies' walls is not to create a boundary between interior and exterior, or to support a roof above. Rather their tectonic gesture is the definition of a limit between discreet spaces, in order to articulate space as such. The minimized width reduces their presence as a mass and transforms them into the lines of a notational system of tectonic conditions.

Contact and interaction of these lines create an intricate masonry system. Two walls that connect in a perpendicular condition are sufficient to create a stiff structure. Pressure that is applied to the surface of one wall is resisted by the reaction of the other wall. The resulting plan of an L or T can be understood as an incomplete rectangular enclosure. Mies works with an arrangement of such incomplete enclosures. Although the plan of the Brick Country House may seem open, many wall configurations look like fragments of common masonry structures. The plan thus represents an open and dynamic space, while maintaining the traditional logic of brick construction.

All the walls in the plan relate to each other in a refined system of alignments. Even small partitions that seem at first glance detached from the overall arrangement align structurally. The strong rectangular framing of the plan suggests a continuity of this spatial concept even beyond the edge of the drawing. At the periphery of the page, a wall is an independent element in the landscape. As it approaches the center, it starts to bear a part of the roof above. But this does not change the thickness of the wall, which remains consistent throughout the plan, making all of its parts potential load bearing structure.

Both the house and the landscape around it are designed as an abstract garden of architectural elements that enclose space within their form. Through the combination of several fragments a dense configuration emerges in the center of the drawing. Here the actual house occurs as a spontaneous result of the tectonic play of the walls. This central

1.6 The plan of the Brick Country House from 1960 shows the bond with which each single brick fits into the logic of the plan.

zone is outlined with a continuous perimeter that notates the climatically sheltered indoor zone. Glazing extends the centerlines of certain walls in strategic locations to create an interior. These glazed openings appear as neither windows nor doors, but as a continuation of the wall in a different material across a gap in the configuration.

BRICK BONDS

The original plan of the project did not survive the war, and was redrawn in the 1960s. The new version shows with great detail and accuracy the *bond* with which each single brick fits into the logic of the plan. During the time in which this drawing was produced Mies worked on the Dominion Center in Toronto and the Gallery of the Twentieth Century in Berlin. He wanted to reexamine a project that he originally proposed in the 1920s, forty years after its initial design.

It seems likely that the new plan drawing represents a point of inspiration for Mies in his work on the Dominion Center towers. Resembling two varied bricks standing in the landscape, the towers continue the investigation of brick proportion and assembly. Understood in this way, they stand on their short end, or *head*, and lift a form that usually lies flat into an upright position. The charged space between the two bricks implies a bond. This makes the towers appear as component elements out of which the emerging city can be constructed. Again, like the Brick Country House, the Dominion Center is a brick still life composed in architectural scale.

Brick Masonry / Case Study

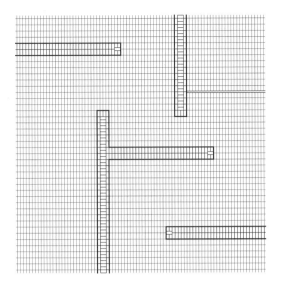

The new plan of the Brick Country House expresses the attempt to fully resolve the design concept both technologically and aesthetically within the logic of bricklaying. Its degree of detail surpasses the earlier drawing from the 1920s. A close look at the plan shows how much care was put into elaborating the different ways in which the bricks are arranged. Here it becomes apparent how the whole house displays the concept of pieces put together to something greater throughout all scales. Architectural plan and construction detail share the theme of interacting and interconnecting pieces, in which the properties of bricklaying are artistically embraced. The spaces of the house are like the brick patterns in the walls themselves: They are rectangular volumes that are connected in

1.7 Dominion Center: bricks in the landscape
1.8 All visible bricks point in the same direction, although the direction of the wall changes. The wall is not conceived as a linear element, but instead as a part of a potentially continuous field of bricks that extends across the entire plan.

various conditions and read as a tectonic elaboration of the bond logic. The plan represents the core idea of a construction method and uses it holistically for the design concept of a house. This constitutes Mies' modern brick architecture: specific to brick, yet independent of traditional stylistic features such as heavy walls, arches or vaults.

The brick arrangement in the walls changes depending on their orientation: All the walls running east-west are composed of two bricks perpendicular to the direction of the wall, while the walls in north-south orientation consist of one brick perpendicular to the direction of the wall sandwiched between two bricks that are parallel to the wall on either side. All the visually exposed bricks follow the same direction: in a wall that goes east-west the surface is composed of all *headers* that show the short side of the brick. When this wall turns ninety degrees the orientation of the bricks remains the same despite this change of direction. As a result the face of the wall now consists only of *stretchers*, the long side of the brick. All bricks point in the same direction, although the direction of the wall changes. This means that the wall is not conceived as a linear element, but instead as a part of a potentially continuous brick volume that extends across the entire plan and is cut into the shape of an L, a T, or a simple line.

ENGLISH BOND
The plan is a conceptual cut through the structure on one horizontal plane. Were the plan cut three inches higher, all bricks would turn ninety degrees.

This *course alternation* changes the direction of the continuous brick texture throughout the house from course to course; the field of bricks switches back and forth between the two cardinal directions established by the walls in the overall plan. Such a concept requires courses that show exclusively either stretchers or headers on the wall surface, which the *English bond* provides. Unlike the *Flemish bond*, each course is distinct from the next, and shows the brick in only one orientation. In this sense the English bond is the most "existentialist" of all bonds and provides Mies with a bricklaying logic that informs the abstract aesthetic concept of the Brick Country House.

Axonometric details show how the English bond turns from course to course in L and T shaped wall connections. A row of stretchers becomes a row of headers when turning a corner. In the next course above, headers continue as stretchers on the other side of the corner respectively. The only exception to the use of the English bond throughout the Brick Country House is the fireplace that is detailed in a *stacked bond*: here the bricks are placed directly on top of each other in a grid without any shift. The hearth is thus another kind of solid and assumes a special role within the overall composition: It seems to be a magical center of gravity around which the elements of the field fluctuate.

Both the perspective and the plan of the house state that brick masonry, despite being an enormously old construction technique, can become flexible and light through the reduced wall thickness that is afforded by the improved material. It can now generate spaces that are airy and dynamic. At the same time the continuous field implied by the use of the *English bond* in the plan suggests that the whole house is a latent masonry solid; the factual lightness is complemented by a conceptual solidity. What is shown on the drawing is only a fleeting moment within an entirely solid creation that is partially brought into physical existence.

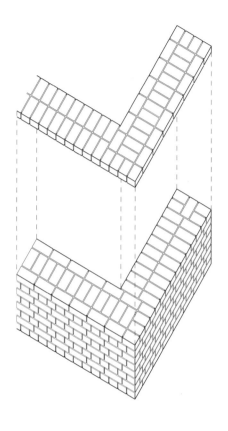
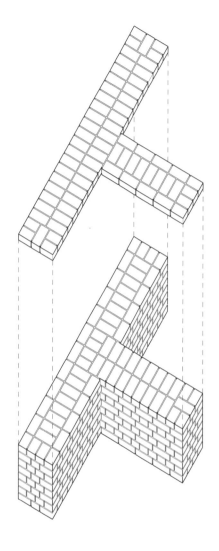

1.9 In the English bond each course is distinct from
the next, and shows the brick in only one orientation
on the wall surface.

Brick Country House

HOUSE WITH THREE COURTYARDS

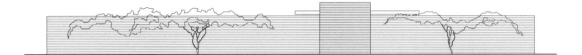

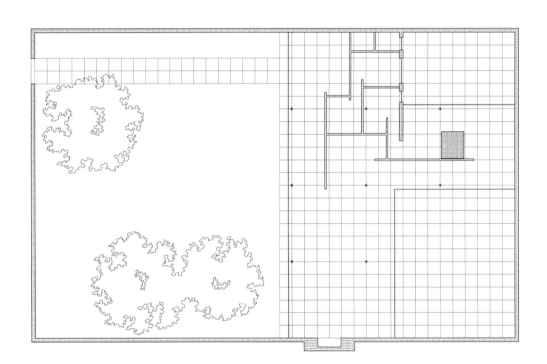

Brick Masonry / Case Study

The House with Three Courtyards designed by Ludwig Mies van der Rohe in 1934 seems at first glance to be a more traditional masonry building than the Brick Country House. It is enclosed by a continuous perimeter wall, which creates an elegantly proportioned architectural volume. All the elements of the house are contained within this static and solid enclosure. Yet its interior contains a dynamic play of elements, creating an intricate arrangement of spaces that seem to be in a constant state of flow.

Three courtyards with similar proportion and different sizes appear to be the main rooms of the house. The largest of them is a garden with three trees, while two smaller ones are paved like the indoors of the house. The interior has an irregular shape and reads as a residual space between the rectangular courtyards. It appears to be a part of the continuously paved ground surface that is also sheltered from above. The only indication of a boundary between exterior and interior is the roof with its T-shaped plan and the glass screens that follow its contour. The continuous perimeter wall of the house defines the outer edge of all three courtyards. This suggests a potential expansion of the spatial organization beyond its current limit: If the enclosure became wider, the courtyards would simply expand outwards without affecting the architectural logic of the plan.

1.10 The plan the House with Three Courtyards is a synthetic cubist composition drawn from various architectural elements that are framed by a rectangular perimeter wall. Mies uses this wall as the same kind of framing device established by the picture frame in the Brick Country House plan.

PICTURE FRAME

This plan is a synthetic cubist composition drawn from various architectural elements that are framed by the outer wall. Pavement textures within the powerfully defined perimeter divide the ground into different zones. Trees and partition walls are the composed objects on these ground surfaces; they repeat the tripartite theme of the three courts. Just like this house features three exterior courtyards within the space of the perimeter wall, the plan of the Brick Country House is also divided into three main zones through the continuous walls that cut across the drawing page. Mies uses the perimeter wall of the House with Three Courtyards as the same kind of framing device established by the picture frame in the Brick Country House plan. With its introverted gesture the courtyard house represents the urban appropriation of a similar spatial arrangement within the intimate cosmos of a rectangular enclosure. Internal partitions between the rooms and courtyards are articulated as different wall types than the perimeter; this underlines the difference between the wall that constitutes the frame, and the walls that partition space within the frame.

It seems obvious that the detailing of the brick bond in this wall needs to be distinctly different from the detailing in the Brick Country House, since it relates to a different architectural theme: it has to support the notion of containment and must distinguish between the inner and outer faces of the wall, rather than implying a spatial continuity across its limit.

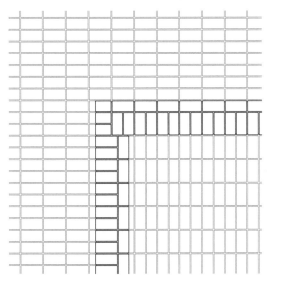
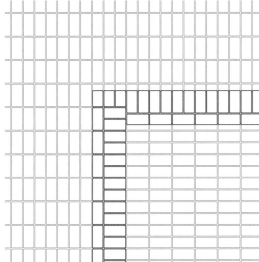

1.11

ONE STRETCHER AND ONE HEADER

Mies uses again an English bond, but this time with a different character and effect: the wall section consists of one header and one stretcher and is therefore only one and a half bricks wide, as opposed to the two bricks wide wall of the Brick Country House. On the interior face of the wall the bricks have a constant orientation, while the bricks on the exterior face of the wall maintain the opposing orientation; this principle is independent of the direction of the wall when it turns. The proportional relationship of one stretcher on one side of the wall to two headers on the other side of the wall is maintained throughout the structure. The only exception appears in the corner where the inside face of the wall transitions from stretchers to headers, and the exterior face transitions from headers to stretchers respectively. In this location Mies introduces *queen closers*, which are bricks cut down to two thirds

of their length. They help to avoid a continuous vertical mortar seam in the very corner, which would appear if the wall turned without any shift in the bond.

The opposing brick direction on either face of the wall manifests the separation between the inside and outside of the house. Inside the wall the contained space is a potentially continuous field of bricks pointing in one direction; this inhabitable volume seems to be cut out of an infinite solid surrounding. From the outside this effect inverts: The bricks point the opposite direction and the wall appears as the outer limit of a homogenous solid block. The entry gate is the single location in which the wall is perforated, and only here the actual thickness of the masonry becomes tangible.

The orientation of the bond that is indicated in Mies' plan refers to the rectangular shape of the house: on the outer face of the wall the short side shows exposed headers, while the long side

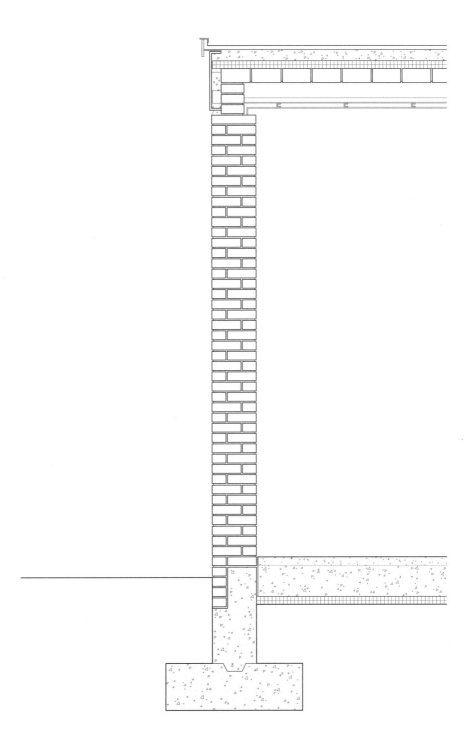

1.11 The opposing brick direction on either face of the wall manifests the separation between the inside and outside of the House with Three Courtyards.
1.12 The steel roof structure is detailed as an independent element that is not integral to the masonry wall below.

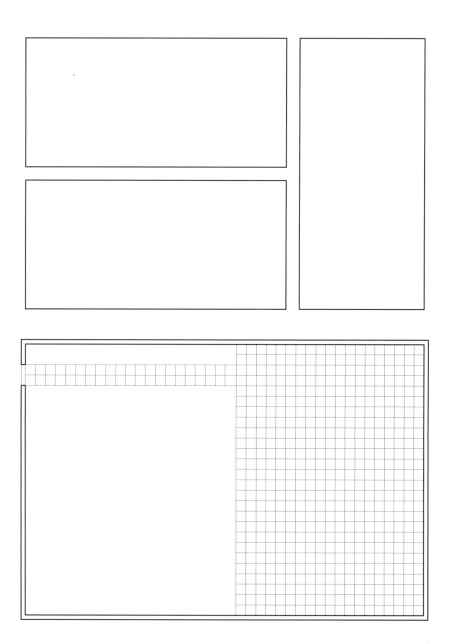

Brick Masonry / Case Study

shows instead stretchers. One simple idea based on proportion unifies the basic construction unit and the overall figure of the house: A single brick is short on one side and long on the other.

Both in plan and elevation the chimney is the only element that projects beyond the outer limit of the wall. It identifies the programmatic nucleus of the house, the center of its inhabitation. In plan it reads as a volumetric extension of the wall. In elevation it divides the total length of the wall into two unequal parts; this suggests yet another reading of the elevation: it consists of a stretcher and a header that are assembled in the English bond as it shows in the cross section of the wall. A close look at the plan logic of the English bond supports this reading: One stretcher meets two headers; the tripartite of these three bricks establishes a rectangle with the same proportions as the overall plan of the house.

ABOVE AND BELOW THE WALL
The roof of the House with Three Court-yards is an additional element sitting on top of the continuous perimeter wall. If the elevation drawing is understood in the scale of a single brick course, the double line of the roof reads proportion-ally as the mortar course on top of it. The roof structure consists of steel beams that are connected to a steel channel running along the top of the wall. It is detailed as an independent element that

1.13 Three bricks laid in the English bond establish a rectangle with the same proportions as the overall plan of the house.

is not integral to the masonry below. Much like the Brick Country House, the wall is primarily a vertical member that has a complete logic in itself. Its thickness does not change depending on whether or not it carries a portion of the roof. The roof could potentially appear in any location and eventually cover the entire enclosure without transforming the tectonic principles of the house.

The brickwork of the wall sits on a concrete *strip footing* that is notched in to receive the outer brick layer; this continuation of the brick below the ground line prevents the concrete foot-ing from becoming exposed through irregularities in the terrain. Inside the wall the floor tiles sit on a concrete slab on grade that connects to the footing. They have the same thickness as one brick, supporting the idea that the space of the house is carved out of a solid masonry volume.

Both the Brick Country House and the House with Three Courtyards trans-form the gestalt of brick architecture without rejecting its traditional charac-teristics. The technological advancement of extruded and wire cut brick provokes a masonry that is less massive and more flexible. However, it does not suggest a different structural application of the material. The houses are still com-posed of traditionally constructed walls. Mies explores the internal assembly of these walls, as well as their external arrangement. They establish an archi-tecture that embraces the great craft of bricklaying, while producing radically abstract spaces.

REINFORCED MASONRY

The invention of reinforced masonry strongly impacted the structural assembly of the brick wall. Masonry structures were traditionally designed only for vertical force flow in which load is transferred downwards in compression. Each individual brick was produced to have as much compressive strength as possible and the brick wall as a whole was bonded and mortared rigidly to resist vertical loads. As a result it could never bend to absorb horizontal forces and cracked easily when stressed in tension. Its composite assembly of small individual units disintegrated under irregularly applied force, no matter how bonding techniques and mortar created a unified structure. In reinforced brick masonry, steel bars are embedded between the bricks to increase ductility. They create a continuous skeleton that ties the individual units together in a powerful way. This technology was pioneered in the second half of the nineteenth century in the Asian colonies of the British Empire in order to make masonry buildings earthquake resistant.

The reinforcement is typically applied in two ways: a grid of reinforcement bars can be sandwiched in a vertical cavity between two parallel wythes of masonry. Once the reinforcement is in place, the cavity is filled with mortar, consolidating the composite wall section into a single piece of structure. Another way of connecting two independently standing masonry wythes is to embed a wire mesh in every other horizontal mortar bed.

Reinforced masonry transformed the bricklaying technique radically. The purpose of all traditional bonds is to assemble the individual bricks into a single structural member. Each bond is a system of laying bricks parallel and perpendicular to the wall in repetitive patterns to generate a continuously interlocking section. This can now be accomplished by the metal reinforcing. No headers that connect into the depth of the wall are required. Two independent wythes are composed solely of stretchers that are shifted horizontally between courses. This *running bond* prevents the vertical alignment of the structurally weaker joints in the elevation of the wall.

The steel reinforcing introduces a new kind of structural unity and tensile strength to brick construction. It eliminates traditional restrictions of the old building method by substituting the bond with the insertion of a new material. The embedded steel allows for unconventional masonry forms that provoke new brick architectures for the twentieth century. These advancements come at a price, the loss of the bricklaying craft. The bonding techniques that were once indispensable and defined the character and beauty of brickwork for centuries were rendered obsolete.

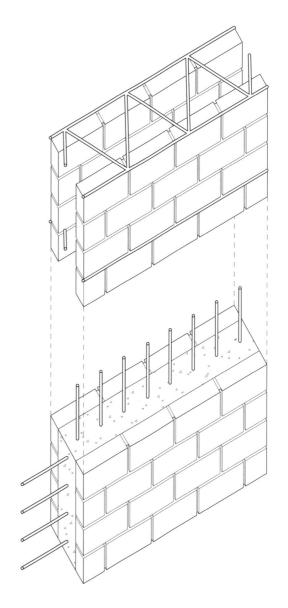

1.14 In reinforced masonry, steel bars connect independent layers of masonry.

1.15 The masonry wythes can be held together with metal ties that span across a cavity (top), or reinforcing bars are embedded in a mortar fill (bottom).

Reinforced Masonry

MONUMENT TO ROSA LUXEMBURG AND KARL LIEBKNECHT

The Monument to Rosa Luxemburg and Karl Liebknecht by Ludwig Mies van der Rohe demonstrates the "liberation of the brick" through steel reinforcing. It was built in Berlin in 1926 and demolished in 1933. The monument only existed for a short period of time and was not published widely in the Post War Era because of its political program: Rosa Luxemburg and Karl Liebknecht were leaders of the social democrats in Germany after the end of WWI. They were supporters of the Spartacist Uprising, a general strike accompanied by armed battles in 1918-19. Members of the Freikorps right wing militia assassinated both of them. The art historian Eduard Fuchs asked Mies to build a monument for them and all the other activists that had been killed during the turmoil of the German Revolution.

BRICK WALL

The artistic leitmotiv of the monument is a wall in its most grim of all possible uses, an execution wall. Many of the revolutionaries were court-marshaled and shot to death by firing squads. The frontal experience of this monument relates to what it stands for: both the execution wall and those condemned to death face the same direction. This violence is embedded in the material out of which the wall is assembled. Mies constructed it entirely out of *rubble bricks* that were salvaged from demolished buildings. They give the wall a coarse appearance suggesting traces of force and impact.

The visually prominent cantilevers could not have been constructed in a conventional masonry bond: the bricks would fall down without additional support. In the monument the dynamic arrangement of shapes is only possible because of the reinforcing steel holding the bricks in place. Mies literally turns brick architecture upside down: a *header course*, a continuous band of vertical headers, would traditionally be used as a coping on the top of a wall. Now it appears at the bottom of each articulate masonry volume as a brick soffit that encloses the cantilevering volumes from below. Such a condition did not exist in traditional masonry construction.

The proportions of the individual masses that compose the wall are examples of all the different shapes a brick can have. An early concept sketch shows that the project is a study of different rectangular elements that are stacked on top of each other. It seems to be a composition of proportions that are all derived from *stretchers*, supporting the visually horizontal gesture of the wall. They are all variations of a horizontal rectilinear prism. At the top of the wall a slim volume resembles a *Roman brick*; other proportions can be found across the elevation of the monument.

1.16 Mies constructed the Monument to Rosa
Luxemburg and Karl Liebknecht from rubble bricks
that were salvaged from demolished buildings. They
give the wall a coarse appearance suggesting traces
of force and impact.

Monument to Rosa Luxemburg and Karl Liebknecht

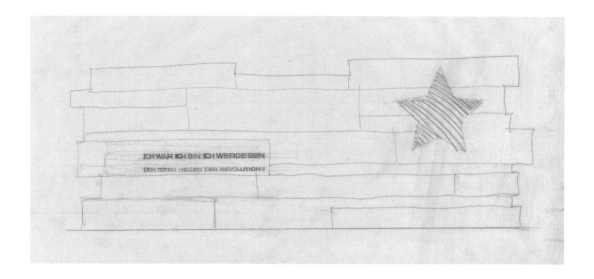

In the site plan of Friedrichsfelde cemetery Mies represents the monument as a single brick that is situated on the ground. It resembles a body that is laid to rest, a tombstone fashioned out of one long brick that memorizes the departed members of the movement buried on either side of it. The alignment of their rectangular graves in plan acknowledges the header courses in the brickwork of the cantilevers. As in the elevation of the House with Three Courtyards, the proportions of a single brick inform the architectural concept throughout all scales. The red color in the drawing associates the materiality of the wall with the political movement that the monument represents. This fusion of physical structure and symbolic metaphor is also emphasized in the elevation: Differently proportioned rectangular shapes convey the idea of diverse components creating a unified composition. The autonomy of this arrangement embodies the voluntary unity of the dynamic members of the movement that is symbolized in the monument: Every individual brick is an important component. As groups they form larger volumetric units, which are slightly different from each other but all follow the same trajectory. Together these volumes constitute a powerful whole, a form that represents the movement in its entirety.

BRICK FLAG

Mies also designed the stainless steel hammer and sickle that was manufactured by the Krupp steel company. At that time Krupp was producing the steel panels for the Chrysler Building in New York. They refused to make the entire hammer and sickle because of the left wing content, yet they provided the individual pieces out of which Mies put the star together himself. The overall proportions of the wall resemble those of a banner and the steel symbol makes it clear that the monument is a sculptural abstraction of a moving Soviet flag. The absence of color in the elevation on a sheet that renders plan and section in

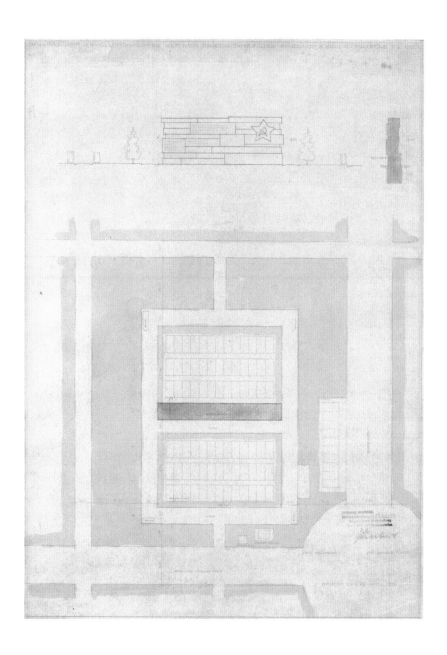

1.17 Differently proportioned rectangular shapes
refer to different brick types and convey the idea of
diverse components creating a unified composition.
1.18 In the site plan Mies represents the monument
as a single brick that is situated on the ground. The
red color in the drawing associates the materiality
of the wall with the political movement that the monument represents.

Monument to Rosa Luxemburg and Karl Liebknecht

red supports this reading: If the elevation was drawn in red the direct formal reference to the flag would have been too obvious. This also explains why the drawn elevations do not show the flagpole that was mounted on the monument to raise an actual flag. In the photos that show this flagpole it never carries a flag. The wall itself *is* the flag! In the Soviet banner as we know it the hammer and sickle are in the upper corner of the hoist end. Mies places the symbol in the upper right corner of the monument, because the fly end of his brick flag is waving towards the left. The monument thus embodies a direct verbal reference to the life of Rosa Luxemburg. She had been the founder of *Die Rote Fahne (The Red Flag)*, the central organ of left wing revolutionaries in Germany.

The flag concept confronts Mies with a tectonic problem: a flag does not stand, but a brick wall has to. He finesses this contradiction with plantings over the graves that cover the bottom of the wall. For the same reason there is no indication of terrain in the section drawing. The datum of ground surface can only be understood through the coordination of the section with the elevation on the drawing page. In itself the section does not indicate a gravitational logic as the bottom of the wall looks identical to its top. The section wants to be understood as a cut through a waving flag. In certain heights it shows alignments between the cantilevering volumes on either of its sides, unifying the entire width of the wall into a stack of shifting units. Yet in other heights, a different shift on either face contradicts this principle, making the width of the wall expand and contract in a dynamic flow.

CORBELLED STEP

The famous corner view of the monument shows a charming masonry detail: a shallow brick plinth not much higher than a single masonry course extends out from underneath the wall and frames the planted graveyard. Close to the wall a small outward projection of the plinth invites the visitor to enter the monument. A second step leads further up to a platform that is the lowest rectangular volume of the wall and allows the visitor to actually inhabit this architecture. This second step is attached to the face of the platform edge and has the dimension of one single brick perpendicular to the wall. It is constructed with the traditional masonry technique of *corbelling*: a brick projects out from within the bond; another brick on top of it projects a little further still to generate a more substantial dimension. Historically this technique was used to create supports for beams on the face of a brick wall. As a traditional masonry projection used in modern architecture it is just big enough for a single step.

The lightness inherent in a cantilever is absent in the heavy appearance of the monument. Its rusticated brick surface and the strong shadows under its projecting volumes make it clear, that

1.19   The monument transforms masonry into a freely moving composition of prismatic shapes. They cantilever into space to establish a new type of brick architecture. A corbelled step that projects out from the base invites the visitor of the cemetery to inhabit the monument.

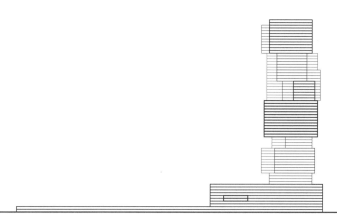

it does not deny gravity. At the same time its expressive form is also full of movement, combining heaviness with dynamic force. Mies works within and against the substance of the brick: Throughout the wall the running bond pattern is very regular, with the same header courses appearing systematically under every volume. But the overall gestalt of the monument transcends masonry into a freely moving composition of prismatic shapes. They project out from the wall and cantilever into space to establish a new type of brick architecture that never existed before.

CONSTRUCTION, DESTRUCTION, RECONSTRUCTION

Mies uses the design of this monument as an opportunity to explore the artistic potential of a reinforced brick masonry wall. He creates an architectural form that can convey symbolic content without becoming pictorial. The monument is dedicated to all the killed activists of the Spartacist Uprising, stated on the metal inscription that was originally mounted on the wall: *Ich war ich bin ich werde sein – den toten Helden der Revolution (I was I am I will be – to the fallen heroes of the revolution)*. The two lines of this inscription form a rectangle that has the exact proportions of the single red brick representing the project in the site plan. The rubble construction technique may be seen as a subtle yet hopeful homage to a failed German Revolution: The monument is conceptually put together from the collected pieces of its ruins. Each single rubble brick in the project incorporates the three conditions of existence that are stated in the dedication. They were first assembled in a building, but reduced to rubble in demolition. Reassembled and reinforced by steel, they became a monument carrying its own history of construction, destruction, and reconstruction.

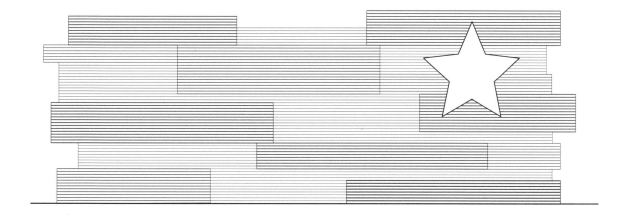

1.20, 1.21 The overall proportions of the wall resemble those of a banner and the steel symbol makes it clear that the monument is a sculptural abstraction of a moving Soviet flag.

Monument to Rosa Luxemburg and Karl Liebknecht

HOUSE IN MUURATSALO

The weekend house in Muuratsalo by Elissa and Alvar Aalto completed in 1953 is another beautiful exploration of reinforced brick masonry. The Aaltos built this vacation home for their own use in the archipelago of the Finnish lakes; it is the last building that Alvar Aalto worked on.

In plan the house reads as an autonomous prism that is placed in the landscape. Two L-shaped walls frame a square courtyard that sits within a square house. One L is a freestanding brick wall; the other is the home's interior, contained within the thickness of the perimeter wall itself. The width of this inhabitable masonry volume is determined by the span of the wooden beams in the roof.

The introverted courtyard typology serves as an architectural leitmotiv for this habitat in the wilderness. Its introversion is complemented by distinct openings to the outside that establish a powerful dialogue with the surrounding landscape. The freestanding brick walls of the courtyard have two openings into the forest: an irregularly shaped window on one side, and a gate on the other. Both are formulated as missing parts of the perimeter that do not want to appear as perforations in the wall's surface. A full segment of the wall is removed from top to bottom to establish the gate. The window is a partial opening also, starting at the top of the wall but not extending to the ground, and is screened with a series of vertical wooden planks. It thus reads as a permeable zone of the wall that is in the process of transition from a fully closed surface to an entirely open condition.

1.22 In plan the house in Muuratsalo reads as an autonomous prism that is placed in the landscape.
1.23 Distinct openings in the perimeter wall establish a powerful dialogue between the introverted courtyard and the surrounding landscape.

House in Muuratsalo

SHED ROOF

The wall with the screened opening that faces the lake is the highest side of the house. From here the roof slopes down towards the interior bedrooms in an angle that resembles the steepness of the topography but slopes the opposite way. Over the interior the roof needs to be tilted to accommodate drainage. The coping of the freestanding brick walls follows that same angle and reads as a slim extension of the roof surface. As a result one slanted plane cuts across the entire building and creates an upper limit in which all vertical elements terminate. This common plane integrates rooms and freestanding walls into a single prismatic volume. A simple shed roof becomes an abstract feature that gives the house the character of a solid volume that has a courtyard space carved out of it.

WALL WITH TWO DIFFERENT SIDES

The experimental use of reinforced masonry in this house demonstrates the separation of a brick wall into two vertical layers that are constructed as independent wythes, which are connected with steel. The two sides of the wall do not have to be coordinated in a common bond and the cavity between them can be filled with mortar or insulating material. This new type of wall is conceptually one of the most complex architectural developments of the twentieth century: What is tectonically perceived as one wall is in fact a pair of walls, an outer and an inner layer running parallel to each other. Although the wall is still used as a contiguous spatial boundary between interior and exterior, it can have a dramatically different appearance on either of its sides, establishing two distinct identities of the same architectural element.

1.24, 1.25 The wall has a dramatically different appearance on either of its sides, establishing two distinct identities of the same architectural element. **1.26** The two identities of the same wall represent the opposition between an artificial and a natural appearance of brick.

House in Muuratsalo

Aalto explores this tectonic problem in the drastically different treatment of the outer and inner wall faces. They represent the opposition between an artificial and a natural appearance of brick. On the outside the wall is skim-coated and painted white, giving the house the semblance of a prismatic volume standing in the forest. From afar the architectonic form is recognized as a sign of human inhabitation: It is an abstract figure that sits in the landscape. On the other side the untreated brick of the inner wall faces blends visually with the natural surroundings: Standing in the courtyard the inhabitant believes that the house is an integral part of the landscape. The forest around the house appears to be a continuation of the courtyard and is thus the inhabitable space of this architecture as much as the courtyard itself. This contradicts the introverted gesture of the courtyard type and turns the house inside out. The gate is again the one place in which the thickness of the wall becomes tangible. Within the threshold, the inner and outer layers are experienced simultaneously.

BRICK PATCHWORK

A patchwork of various brick types and patterns covers the inside face of the courtyard wall. Three different segments of this patchwork relate proportionally to the three horizontal zones that comprise the wall: A base portion, the height of the inhabitable rooms, and the triangular upward extension that accommodates the tilted roof. The tilted roof edge supports the expressive character of the brick patchwork. Rather than culminating in a horizontal course of bricks continuous with the patchwork pattern, the roof slices the wall in an angle that is in contrast to the overall texture. This makes the brick surface appear as an abstract and potentially continuous pattern, rather than as a construction assembly carrying a roof. The brick patchwork looks like a cubist landscape painting that merges visually with the actual landscape beyond the walls. The reference to nature is also maintained in the proportions and textures of the openings. A gradual transition between the architectural and natural landscape is evident in the screened window of the courtyard wall. Its bottom sill has an erratic profile that looks like the rocky ground of the hilltop terrain; its vertical screening resembles the thin trees of the surrounding forest, blurring the house's boundary from inside the courtyard. In contrast to this visual collapse between house and forest from the inside out, the white wall surface establishes a visually drastic boundary from the outside in.

In comparison with the cantilevers of the Monument to Rosa Luxemburg and Karl Liebknecht, the varying brick arrangements in Aalto's reinforced masonry emphasize surface texture rather than volumetrics. Since the bricks are not tied together in actual bonds, the courtyard elevation shows only non-structural stacked bonds or running bonds. Headers can only fit into this single layer of brick when they project out into the courtyard, rather than into the depth of the wall. This is beautifully demonstrated in the rectangular projections from the wall surface that surround

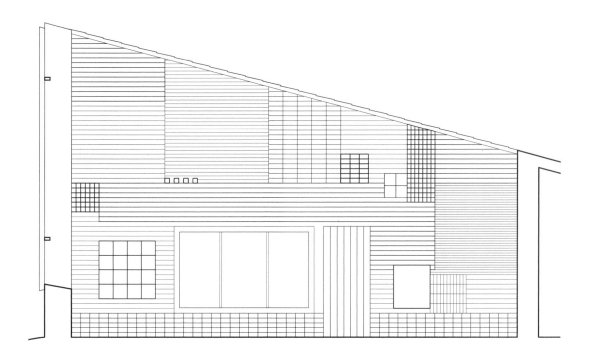

1.27, 1.28 (overleaf) A patchwork of various
brick types and patterns covers the inside face
of the courtyard wall.

House in Muuratsalo

the window. They can be understood as displaced masonry patches that once filled the wall in the locations of its current openings. The L-shaped opening into the living room, consisting of windows and an entry door becomes a part of the patchwork from which certain pieces have been removed. It appears as an integral part of the wall texture, rather than as a hole in its surface. The door features a vertical wooden screen, similar to the erratic opening in the courtyard wall. This screen diffuses the appearance of a traditional door, and makes it appear to be yet another patch.

EXPANSION JOINT

The patchwork experiment of the courtyard walls points to yet another aspect of reinforced brick masonry construction: the joint. Layered cavity walls expand and contract, requiring preventative design to avert cracking under thermally induced stress. Expansion joints divide the wall surface in regular intervals and allow the individual wythes to move independently. The joints become necessary in reinforced masonry wall construction, because the wall consists of thin layers with a small thermal mass. Expansion and contraction occur much faster than in traditional brick construction, where a large thermal mass retards the effects of changing temperatures. Since the wall is not a unified solid anymore, the individual layers are exposed to different temperatures and move differently.

Expansion joints are usually regarded as an aesthetic disadvantage of contemporary brick construction, because they compromise the solid and regular appearance of a wall's surface. They are usually downplayed, yet in any case remain visible. In the house in Muuratsalo this technological necessity is celebrated as an artistic opportunity: the joint is articulated as a seam, a demarcation that is used to underline the character of the various patterns that it separates within the brick patchwork. Together all the seams establish a larger texture across the entire surface of the wall.

FIRE-PLACE

Brick can be used effectively as a paving material; this allows the Aaltos to continue their brick patchwork in the horizontal surface of the courtyard as well. Its plan reads as a floor elevation that also features a window to the natural surroundings of the house, this time down into the ground: the fireplace. Like in Mies' houses the fireplace establishes a programmatic nucleus, in reference to which all the other elements are organized. In Muuratsalo the fireplace is a square brick frame around an exposed piece of earth. It repeats the square courtyard wall in a smaller scale and thus becomes a conceptual condensation of the entire house. A hearth situated in the landscape is symbolically the most primal act of human inhabitation. The *fire-place* marks the domain of man in the wilderness and the house around it represents its enlarged rendition.

The similarity of wall and floor makes the brick textures appear as abstract surfaces that are not derived from a structural logic. A traditional use of masonry would obviously result in

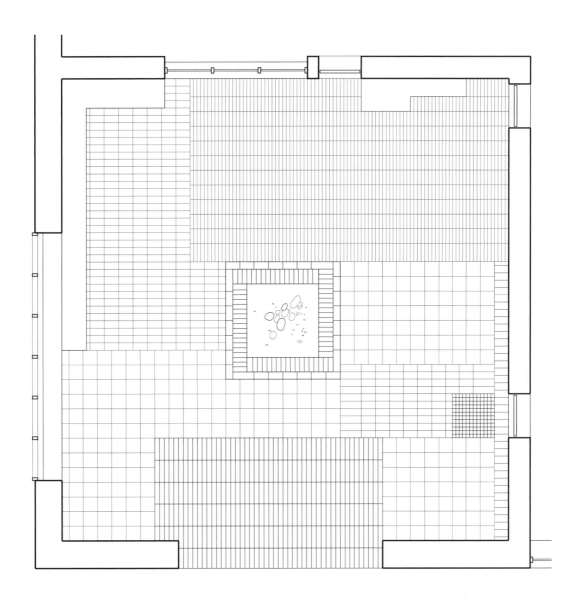

1.29 The fireplace in the center of the brick paving
repeats the square courtyard wall in a smaller scale
and becomes a conceptual condensation of the
entire house.

House in Muuratsalo

a different treatment of wall and floor. The three-dimensional application of the same patterns underlines the aesthetic concept of the house: it is a brick volume cut open and then carved out; through the carving the internal texture of its substance is revealed.

AURA

All together the house in Muuratsalo has a very humble character. Full of humor, it is constructed with modest means and shows a playful attitude towards building a home in the world. At the same time all its features are conceptually precise and rigorous. The artistic concept transcends its simple components into a meaningful work of architecture.

Walter Benjamin writes about art as it pertains to aura. According to him a traditional art piece radiates an aura based on the experience of a solemn distance that it provokes in us. "A distance as close as it can be." The artwork points to the afar while being close to us and attached to our inner self. Benjamin's approach to art is beautifully exemplified in this house. Within the introverted enclosure of the courtyard our attention is turned out towards the surrounding wilderness. Yet, through the artistic treatment of the walls, which read as a reproduction of nature, the courtyard space is the space of the forest turned outside in. Reciprocally, the space of the forest is the space of the house turned inside out. The inhabitant of the house seeks refuge in the afar of the court, which is surrounded by the infinite closeness of the wood.

ARTS VERSUS CRAFTS

Both Mies' and Aalto's brick masonry projects gain their architectural expression and spatial experience from a confrontation of history and innovation. The construction parameters and spatial typologies of traditional brick masonry encounter abstract concepts that are provoked by technological novelties. The projects are architectural essays about the material and its application, rather than structures built from within the traditional craft that the material memorizes. The modernist approach to masonry is to use the new technological possibilities as a means to reevaluate its traditional qualities. Their architectural potential can only prevail in modern construction, when explored in an unconventional manner. Industrialization has dissociated the architect from the traditional craft of brick building and thereby eliminated its historical aura. The only way for the architect to reassign a new aura to the transformed construction method is to employ an artistic concept, and to build an essay about the losses and gains of the construction matter in the technological age.

1.30 A home in the world.

THIN SHELL CONCRETE

LOS MANANTIALES RESTAURANT

MEISO NO MORI MUNICIPAL FUNERAL HALL

THIN SHELL CONCRETE

Throughout the history of architecture, large support free interiors were created using compressive structures such as masonry domes and vaults, or tensile structures such as tents made from thin textile membranes. Their graceful lines embody a history of enclosing space that ranges from the monumental to the nomadic. The form of these structures assumes the curved shape of an upward parabola or a downward catenary respectively.

The advent of thin concrete shell technology in the twentieth century allowed for structures that can assume both of these types of curves using a single material, creating a smooth monolithic surface. Reinforced concrete can be poured into any desired size and shape. As a composite material that combines the properties of stone and steel, it can be stressed both in compression and in tension; thus it can incorporate the curvatures of both vaulted and hanging structures in a single shape. This shape can be distilled into a thin membrane due to the strength of the material, representing nothing but the pure flow of forces through space. Lightweight roofs built in this manner seem to defy gravity in their effortless span. Their ability to cover large distances with little material redefined the notion of sheltered space in modernity.

Thin Shell Concrete

COMPOUND SURFACE DESIGN

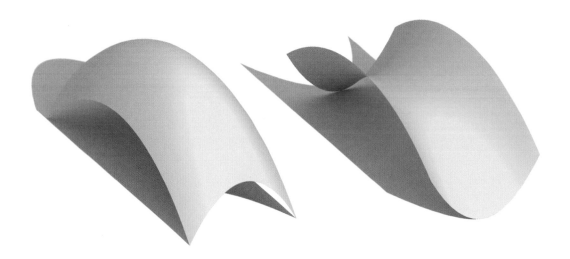

A concrete shell is a structure that acquires strength from its shape rather than from its thickness. Geometry is the main criterion for its design; it defines how a shell will act. Different types of surfaces were developed in the twentieth century, in the effort to reduce the amount of material required to construct a large span. In *simply curved surfaces*, i.e. surfaces that are curved in one direction, the bending of the material performs a crucial part of the structural action. Concrete shells that are based on this kind of geometry, such as the *barrel vault*, need to have a considerable thickness because they must resist *flexural stresses* that impose compression and tension on the top and bottom of the section respectively. *Compound surfaces* that curve in two directions create a stronger shell. Such forms are stiffer, and can resist loads by means of *membrane stresses* that are parallel to the shell

surface in either compression or tension. By distributing these stresses equally in the thickness of the material, the full section of the member becomes structurally active and its thickness can thus be minimized. An egg is a familiar thin shell that illuminates this behavior well: It resists pressure applied uniformly to its surface, though its surface is very thin.

To produce such an optimized shape from reinforced concrete, two kinds of compound surfaces can be used: in *synclastic surfaces*, which are also called *elliptical*, both curvatures go in the same direction, as is the case in a *dome*. In *anticlastic surfaces*, which are also called *hyperbolic*, the main curvatures run in opposite directions, as in a *saddle*. Many of these double curvatures are expensive to build in concrete, because complex formwork has to be constructed from curved elements to pour the liquid material into the desired shape.

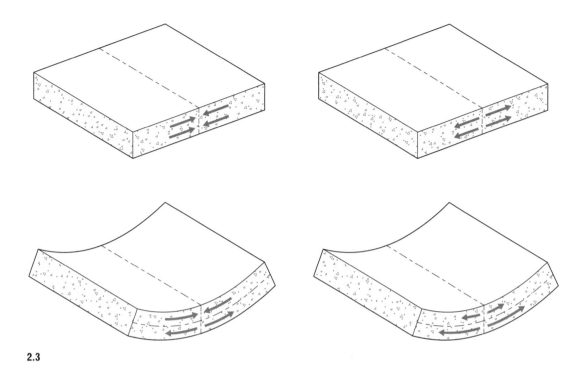

2.3

2.1 (previous page) Thin shell concrete structures can incorporate the curvatures of vaulted masonry and textile membranes in a single material.

2.2 Compound surfaces that curve in two directions create a strong shell. Left: synclastic (elliptical) surface—dome; right: anticlastic (hyperbolic) surface—saddle.

2.3 Membrane stresses that are parallel to the curvature of a shell distribute equally in the thickness of the material. The full section of the member becomes structurally active and its thickness can thus be minimized. Top left: membrane stress—compression; top right: membrane stress—tension. Bottom left: flexural stress—pure bending; bottom right: combined flexural stress and tensile membrane stress.

Compound Surface Design

Warped shapes such as the *conoid*, the *hyperboloid*, and the *hyperbolic paraboloid* are anticlastic surfaces that can be geometrically defined by straight lines that are called *generators*. This means that they are much simpler to construct as concrete shells, because their formwork can be assembled from straight boards. Because of the great stiffness of its double curvature, the hyperbolic paraboloid, also called hypar, is a warped shape that is particularly popular for thin shell construction. It is the translation of an upward parabola over a downward parabola, in which a series of arches placed in linear sequence run along a set of cables that go in the other direction. The hypar is the only warped surface whose geometry is simple enough to permit stress calculation by elementary mathematics. Structures of this shape were both easy to execute and easy to calculate before the computer had replaced the slide rule in engineering. They could be designed without great effort and erected with impressive speed. The simple construction of their formwork in combination with their small amounts of concrete and reinforcing steel created economical large span shells that represent a remarkable fusion of technological efficiency and artistic expression.

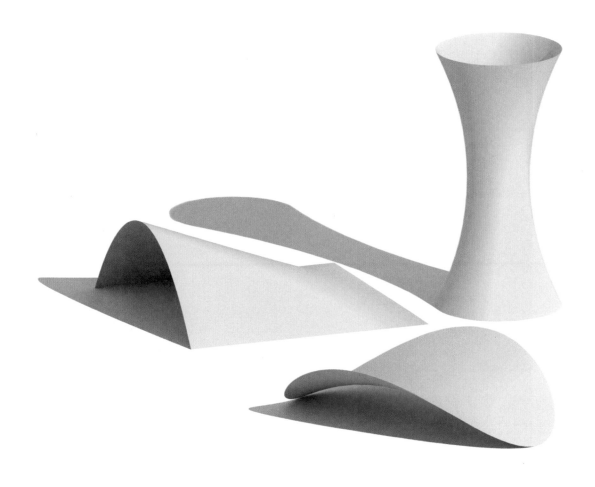

2.4 Warped shapes can be defined geometrically from straight lines. They are easily constructed as concrete shells, because their formwork can be assembled from straight boards. Top left: conoid. Top right: hyperboloid. Bottom: Hyperbolic paraboloid, also called hypar, or saddle.

Compound Surface Design

LOS MANANTIALES RESTAURANT

The Los Manantiales restaurant in Xochilmilco constructed by the architects Joachin and Fernando Alvarez Ordonez in collaboration with the engineer and contractor Felix Candela in 1958 is a concrete shell design that uses the geometric properties of the hypar to arrive at a lightweight roof structure that is both economical and spectacular. The building consists of a single shell that spans over the inside and outside seating of a restaurant in a park near the center of Mexico City. This thin structure looks like a piece of cloth that would blow away were it not held down to the ground by its points of support.

SCULPTURE AND BASE

The restaurant sits on a terrace at the edge of a canal. The terrace acts as a base for the concrete shell, and its formal features are determined as aesthetic counterpoints in support of the art piece that it carries. Its rectangular shape has a strongly architectural appearance, complementing the abstract sculptural gesture of curvilinear structure on top of it. Its vertical faces tilt inwards with a masonry *batter*, a traditional feature of stone architecture emphasizing solidity and heaviness: to ensure the stability of a wall or plinth, the thickness of the masonry increases downwards in

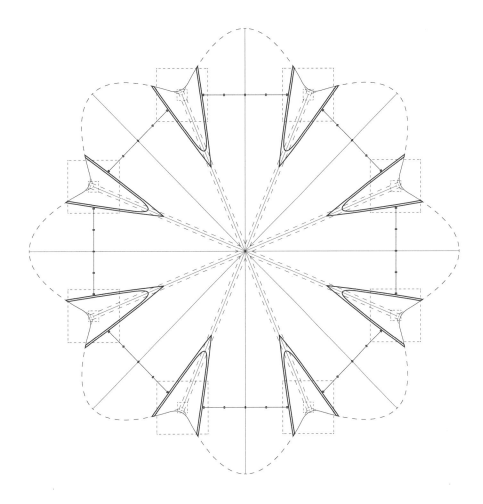

proportion with the load it carries. This visual evidence of the base giving in to gravity contrasts with the dynamic appearance of the shell that seems to be floating above the ground. Just as the heavy rectangular form of the base emphasizes the lightness of the thin concrete curves, its rough texture underlines the smoothness of the concrete surface.

2.5 The design of the restaurant Los Manantiales uses the geometric properties of the hyperbolic paraboloid to arrive at a lightweight roof structure that is both economical and spectacular.
2.6 The concrete shell consists of four intersecting hyperbolic paraboloids that compose an octagonal vault.

The shell is detached not only from the gravitational force of the ground, but also from the historical and cultural implications of the site. Smooth and scale-less, the shape denies any reference to traditional Mexican architectures, or to any known vocabulary of built form. In contrast, the masonry form of the base resembles the remains of an ancient Mayan pyramid rising from the water. The particular design of the base places the shell in its cultural environment and provides the ground-work for a modern Mexican concrete architecture that is built upon Pre-Columbian heritage.

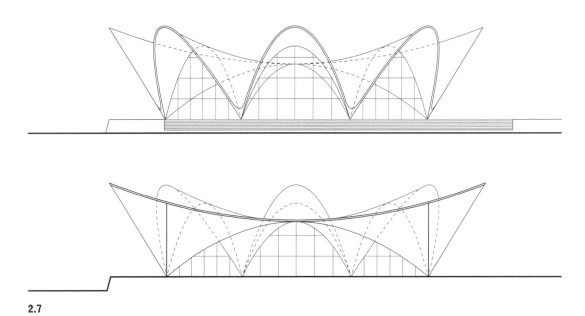

2.7

FORM-WORK

Because of its homogenous appearance, concrete emphasizes the shape into which it is poured. In a thin concrete shell like the Los Manantiales restaurant, this formal emphasis of concrete architecture becomes even stronger because the volume of the material is reduced to a surface that can be as thin as a textile. The scale-less appearance of the material enhances the mathematic concept of the form, resulting in a shape that is open for associations suggested by its pure geometry: floral forms, water waves, or even atmospheric phenomena and musical sounds resonate in its undulating curves.

The lightweight roof of the restaurant is the product of a mathematical operation that is directly rendered as a physical shape. It consists of four intersecting hyperbolic paraboloids that compose an octagonal vault. The engineered optimization of the hypar geometry allows the resulting shell segments to be only

four centimeters thick. The totally symmetrical arrangement of these segments is revealed in the roof plan. Each part is a pure anticlastic compound surface in which the pressure line of the loads coincides exactly with the curve of the cross section, so that there is compression at every point.

Invisible outward tilting planes define the perimeter of the roof, creating prominent arches that cantilever into the water and vegetation of the park. Candela introduces this geometric cutting surface to constrain the potentially infinite hypar shapes to a certain limit. The top of the outward tilt defines the widest diameter of the roof: 42 meters, the equivalent of 139 feet. This feature would not have been possible in a traditional masonry vault. The outward tilt of the shell seems to defy gravity just as radically as the inward tilt of the base celebrates giving in to it. The geometric operations of the shell design result in a curved object that

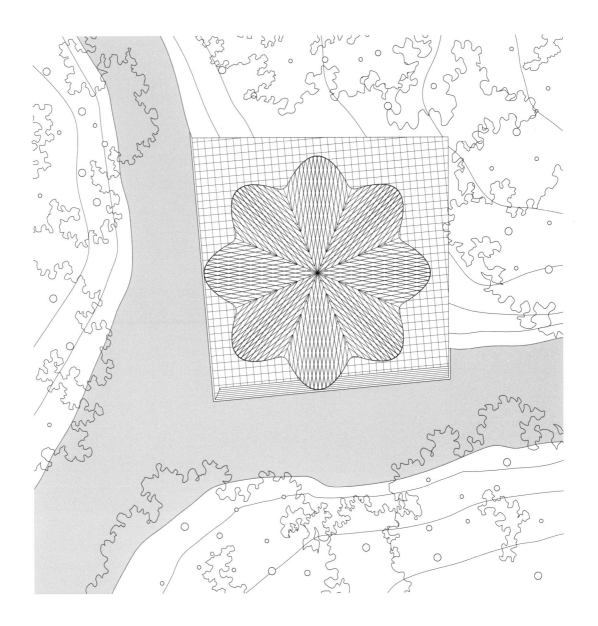

2.7 The roof is a mathematical operation that is directly rendered as a physical shape. The outward tilt of the shell seems to defy gravity just as radically as the inward tilt of the base celebrates giving into it.
2.8 The undulating curves of its pure geometry suggest floral forms, water waves, and musical sounds.

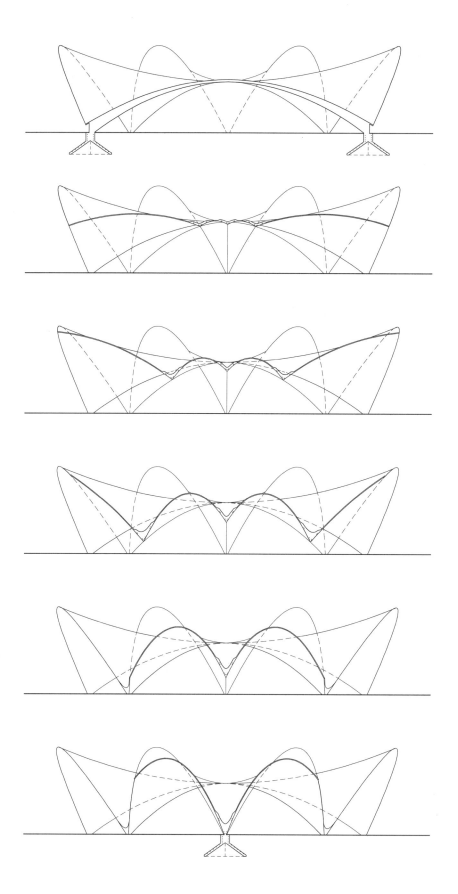

Thin Shell Concrete / Case Study

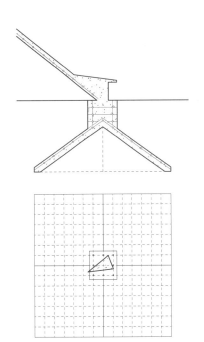

does not allow the form to visually decompose it into constituent parts. There is no tectonic hierarchy of individual members from which this structure was assembled, as can be found, for example, in a post and beam frame. The structure is only comprehensible as a shape in its entirety, and the curvature of this shape is a direct *formulation* of the forces required to create the large span between its supports. For this reason the lightweight concrete roof reads as a visualization of the mathematical rules that represent the curve as such. This relates the aesthetics of the building to natural science and music, disciplines

2.9, 2.10 Curved beams along the intersections of the hypars collect the loads from the shell and carry them to the point supports.
2.11 The footings are made from concrete shell hypars that are assembled to the form of a pyramid.

that use mathematically defined curves to visualize abstract concepts.

GROINED VAULT

The load carried in the shell-surface needs to be channeled to meet the supports. Placing a column under the concrete membrane without funneling the loads towards its location would penetrate the shell just as a needle pierces an egg. In the roof for a centralized building like this restaurant, it would have seemed appropriate to collect the loads from the shell with an edge beam along the perimeter where the roof touches the ground; but this would have disguised the thinness of the shell. In order to avoid this, groins run as curved beams along the intersections of the hypar segments. They collect the loads and carry them to the supports. No edge stresses that would have made an edge beam necessary occur in the

2.12

symmetrical shape of roof. High concentrations of forces occur instead along the groins. The concrete is thickened outwards from four to twelve centimeters and steel reinforcing bars are added. An in-built beam emerges within the homogenous material of the concrete shell.

Candela does not reveal the structural groins to the eye. At the edge of the roof, he pulls a four centimeters thick layer of concrete in front of the supports. It makes the whole concrete structure look as if it was cast in one consistently thin dimension. The extended roof edge comes close to the ground but never actually touches it. This visual suppression of the support makes the whole shell appear to hover—a design feature that further enhances the defiance of gravity.

The curved force flow in a groin vault results in an outward acting thrust at the support; the foundation of such a structure must not only resist vertical but also horizontal forces. But in the lightweight roof of this restaurant, the overhang forces of the outward tilt act opposite to the forces that are flowing down the groin. Thus they reduce the outward horizontal thrust and enhance the structural equilibrium of the shell.

UMBRELLA FOOTINGS
The soft soil conditions of Mexico City would have made a traditional flat concrete footing at the point-supports large and expensive to construct. Candela uses a special foundation technique in which the footings act like umbrellas with a large surface and a little mass. These *umbrella footings* are made from concrete shell sections in the shape of

Thin Shell Concrete / Case Study

2.13

2.14

hypars, which are assembled to a pyramid. This creates a firm shape with a large surface area that distributes the load to the soil while using little construction material. Each footing is a thin concrete shell structure whose shape is designed to distribute load from a point to a large ground surface. They operate the inverse way the roof collects load from a plane and funnels it to a point.

The individual umbrella footings support only vertical loads. A perimeter ring of five steel bars connects them below ground and restrains lateral forces that can occur in an earthquake. Mexico City is a region of high seismic activity. The earthquake force imposed on a structure is proportional to the building weight times ground acceleration; thus the small weight of the restaurant roof makes it resistant against the impact of ground shaking.

2.12 The construction of the umbrella footing exemplifies a use of modern concrete technology that is both pragmatic and poetic.
2.13 Timber shoring provides the substructure for the wooden formwork of the lightweight concrete roof.
2.14 The hypar shape of the formwork for the shell can be assembled from straight wooden planks.

CAST IN PLACE

All the concrete of the restaurant is cast in place directly in its permanent location on site. While the geometric design concept is based on international research accomplishments in reinforced concrete engineering, it is also suitable to the specific environment at Xochimilco and takes into account the construction possibilities and restrictions of the place in which the abstract shape is produced.

The construction of the umbrella footing exemplifies a use of modern concrete technology that is both pragmatic and poetic. For these footings, the ground itself serves as the formwork material. A roughly shaped pyramid is sculpted from the claylike soil with the help of a wooden template. This earth form is then coated with cement mortar to establish the precise shape of the hypar segments, with the help of a pattern of threads representing their generators. Steel bars spaced at eight inches on center are placed on this hard surface and the concrete is patched on top of them. Steel reinforcing sticking out of the top of the pyramid is provided for the connection to the point support, so that groin and footing can be structurally integrated.

Los Manantiales Restaurant

A dense forest of timber shoring provides the substructure for the wooden formwork of the lightweight roof. Its resemblance of a large Polynesian hut put together from wooden sticks gives evidence of the low-tech construction procedures employed to create this high-tech shape. Since a hypar shape can be defined by straight lines, the form for the whole shell can be assembled from straight wooden planks that are placed on top of the shoring. The repetitive geometry of the roof structure may have suggested the use of one repetitive formwork piece that gets used several times for the eight segments of the octagonal vault. But Candela does not use such movable forms for the project and chooses instead to form the entire roof at once. Forming one *petal* of the shell at a time would not have produced the forces necessary to balance each petal against the adjacent ones; the structural integrity of the optimized shape depends on emerging as a whole during construction.

The hypars only need nominal mild steel reinforcing to protect them against small deformations caused by temperature expansion. The reinforcing bars are laid close together so they can support the wet concrete in its place during the *hydration process*, while it is hardening.

Due to the use of a *single formwork* that does not provide a container for the wet concrete, it has to be a relatively dry mix, so it will not run down the sloped surfaces. A thin layer of liquid cement is poured onto the wooden form first to generate a smooth surface for the underside of the shell; then the dry mix is applied. Laborers carry it up the slope in buckets and place it by hand.

After the concrete has hardened, the precise and careful removal of the formwork is an integral part of the operation of building this lightweight roof. It is important to avoid any irregular stresses on the thin concrete member that could occur if the formwork was removed in some parts only while left in place in others. A totally equal stress distribution at any time is essential for the structural performance of such a thin concrete shell. The symmetry of the force flow has to be maintained throughout all phases of construction.

2.15 The steel reinforcing bars are laid close together so they can support the wet concrete in its place during the hydration process.
2.16 Due to the use of a single formwork the concrete has to be a relatively dry mix, so it will not run down the sloped surfaces.

Los Manantiales Restaurant

IN SPITE OF THE CONCRETE

In 1911 the art historian Wilhelm Worringer said that "everything that the Gothic architecture accomplishes, is accomplished in spite of the stone. Its expression is not built upon the matter, but only comes into being through the denial of the matter, only through its dematerialization." The architectural features of Gothic cathedrals derive from the desire to create an interior that is large and uplifting out of as little stone as possible. For this purpose, geometric concepts dissolve the solid and heavy material into a filigree of buttresses and tracery, optimizing the force flow within their curved shape.

Candela's geometric concept for the compressive system of the Los Manantiales restaurant is based on structural principles that had been used in the vaults of Gothic cathedrals hundreds of years earlier. Although this architecture seems formally quite distant, the conceptual approach to materiality is similar. Both architectures are distinguished by the attempt to diffuse the physical presence of their material, in order to arrive at a total exaltation of space. In this sense, Candela's concrete shell is built "in spite of the concrete." As an artificial stone with embedded steel, a solid and heavy gesture would seem to be the most obvious appearance of this material. Instead it is refined into a piece of cloth that "denies its own matter." As with the Gothic cathedrals, the use of geometric concepts minimizes the loads and stresses, to arrive at a vaulted space that is an advanced rendition of its medieval ancestor.

WAVELENGTH

In Candela's vault the filigree masonry of the Gothic cathedral has been replaced by a thin concrete surface. The symmetry of its shape results in an elegant gesture, but also makes the spatial expression of this architecture quite definitive. The structural stiffness of the hypar has the unavoidable consequence of formal rigidity. Its explicit character and reference to natural shapes leaves little room for the observer to explore the building and discover something personal in it that lies beyond its shape. A magical force seems to detach the undulating curves from the ground and gives them a sense of extreme lightness. At the same time these curves appear static and seem to be frozen in a particular moment, without giving the observer the illusion that they will ever be able to change. If the ephemeral appearance of this dematerialized curve is understood as the visualization of a sound, it would be a consistent tone with a fixed wavelength, rather than a playful melody.

Thin Shell Concrete / Case Study

2.17 A magical force seems to detach the undulating concrete curves from the ground and gives them a sense of extreme lightness.

Los Manantiales Restaurant

EVOLUTIONARY STRUCTURAL OPTIMIZATION

The introduction of digital technology in structural computation in the late twentieth century radically widened the range of possible geometries in thin shell design. Irregular curvatures that had always been difficult to calculate could now be analyzed and optimized with the help of the computer. When a designer had previously employed a regular curvature like a hypar or other clearly defined geometric figures for the form of a thin shell, the behavior of the structure could be understood with basic mathematical concepts. Forces and resulting stresses could be calculated and the thickness of the member could be dimensioned accordingly. Since the Renaissance this has been the procedure of designing a large span structure.

In 1992 the engineers Mike Xie and Grant Steven developed the computer aided design method *Evolutionary Structural Optimization (ESO)*. This technique for finding optimal structural shapes is based on the concept of gradually removing inefficient features from the topology of a given shape. ESO inverts the traditional relationship between mathematical concept and building form in the design process. The design begins with an initial sketch by the architect that formulates the artistic intention of the building without any structural considerations. This arbitrary shape is measured and recorded digitally and thus defined in a three-dimensional model of

coordinates. It then undergoes a *sensitivity analysis*, which is a technique for systematically changing the parameters of material, support location etc. in the digital model of the shape to determine the effect of such changes.

This analysis reveals how the shape would perform as a concrete shell. It visualizes the force flow and indicates maximum and minimum stresses as well as the resulting thicknesses of the material. It thereby becomes clear which parts of the shape would perform well structurally and which areas are problematic. In order to improve the zones that perform poorly, the architect can now make any number of adjustments to the initial shape in response to the suggestions provided by the computer. The curvature of the shape can change, its size can be adjusted, its support points can be repositioned, and reinforcing can be added in strategic locations— whatever is necessary to redistribute forces. This transforms not only the structural characteristics but also the spatial configuration of the building and so creates an edit of the outgoing sketch. This second incarnation of the shape undergoes another sensitivity analysis and the process of adjustment is repeated. The final form and curvature of the structure is developed over many such evolutionary cycles that each combine structural analysis and its design response.

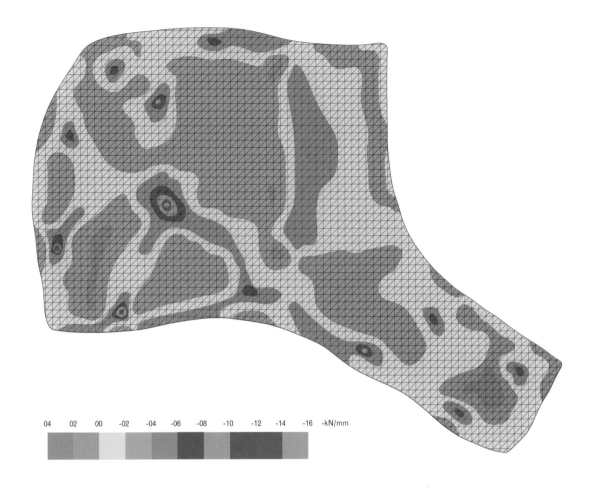

04　02　00　-02　-04　-06　-08　-10　-12　-14　-16　-kN/mm

Evolutionary Structural Optimization is a dialogue between human creativity and digital aid, which are responding to each other back and forth. The design work does not begin with the idea of a specific architectural form, but instead with a spatial intention and a method for how to work towards the form. The actual shape is unknown in the beginning and only discovered over many cycles by

mathematical operations that are considered artistically. While it may look quite different from the initial sketch, the final shape retains the seemingly spontaneous and lively appearance of the first version because it represents its structurally optimized rendition.

2.18 An arbitrary shape can be measured and recorded digitally. The sensitivity analysis reveals how the shape would perform as a concrete shell; it visualizes the force flow and indicates maximum and minimum stresses.

MEISO NO MORI MUNICIPAL
FUNERAL HALL

The funeral hall that Toyo Ito designed in collaboration with the engineer Mutsuro Sasaki is a great example for the use of Evolutionary Structural Optimization. It was completed in 2006 in the small town Kakamigahara in Japan. The curvature of the structure follows the established concept of membrane stress distribution to minimize the thickness of the material; but this concept is realized in a totally free form that goes beyond established geometric figures and is not determined by a preconceived idea for a specific type of span. The resulting architecture presents itself as an undulating concrete roof whose shallow curves transform gradually downwards into slender columns.

The way in which the shell and its supports are arranged in the landscape makes the building appear to move across the site in a gentle flow. A continuous ambulatory under the roof transitions fluently into the exterior space. All programmatic components of the funerary hall are designed as individual building volumes that stand independently in this sheltered space. Their rectilinear forms distinguish them from the white concrete structure above, and their unique scale and materiality reflect their different functions. A ceremonial sequence organizes these volumes on a continuous ground level between the edge of an artificial lake and the foot of a wooded hill. Upon arrival at a covered driveway, an entrance hall gives access to the ambulatory. A series of low wooden compartments that open towards the lake serve as waiting rooms in which the families of the departed gather before the ceremony. The valedictory, the place where the mourners say goodbye to the departed, is housed in a stone clad cube with prominent doors. From here the funerary procession continues to the main hall that is open to the concrete roof above. It faces the wall of the furnace room, which rises to a prominent dome in the shell over the center of the building.

While the shape of the funerary hall is determined by artistic considerations, there is also an emphasis on the structural logic of its form. The columns that stand close together have a shallow span between them; the further they are located apart, the higher the shell rises. In this project these basic rules of vaulting become highly complex, because they occur over irregular support locations dispersed in the terrain. In the application of the ESO method the design of the roof structure began with the artistic intention to span a continuous thin concrete shell over an irregular arrangement of spaces and supports. The formal gesture is best understood as throwing a piece of cloth over the programmatic volumes and establishing certain rules for how this cloth behaves, as follows:

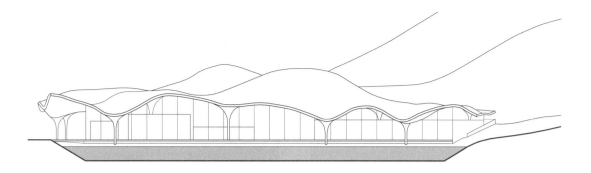

1. It is supported by the walls of the furnace room.

2. It has to float over the valedictory and waiting area.

3. It is supported on columns that can stand on the floor of the interior ambulatory, on top of the interior rooms, and directly on the ground in the exterior.

4. The columns are always located under the low points of the roof. From here the shell springs in an upward bow and spans over to the next support. It can also cantilever along the edge of the building.

5. The columns have an identical shape but they can stand on different elevations and thus support the roof at different heights.

Upon defining the initial gesture for this structure, Ito and Sasaki worked on it for over 150 evolutionary cycles. During this process all the defining parameters could be adjusted. The vertical supports could be moved since the loads did not have to go down in specific locations; the curves of the roof could move up and down freely between them with

2.19 The continuous shell of the funeral hall spans omnidirectional between irregular support locations. It presents itself as the lowest hill of the valley beyond.

reinforcing added in locations of high stress; even the programmatic volumes could change their shape and be relocated as long as their arrangement continued to serve the program.

During such an evolutionary design process the structural analysis may indicate that a certain location in the concrete roof is a weak point. If it is not possible to place a vertical support that would solve the deficiency under this location, then the issue needs to be addressed by a different kind of shell curvature, or the configuration of program around the location needs to be changed, in order to allow for the column. Primal form and location of the components are given, but everything else is flexible.

The resulting building has formal similarities to its 150 generations old ancestor, but is a totally different shape. Its design is not determined by an outgoing assumption in which the architect and engineer say "this is the shape that we want," but instead "this is the *kind* of shape that we want." The structural performance of this kind of shape is gradually improved based on its digital analysis, while maintaining artistic control over how these improvements are implemented. In this procedure, the

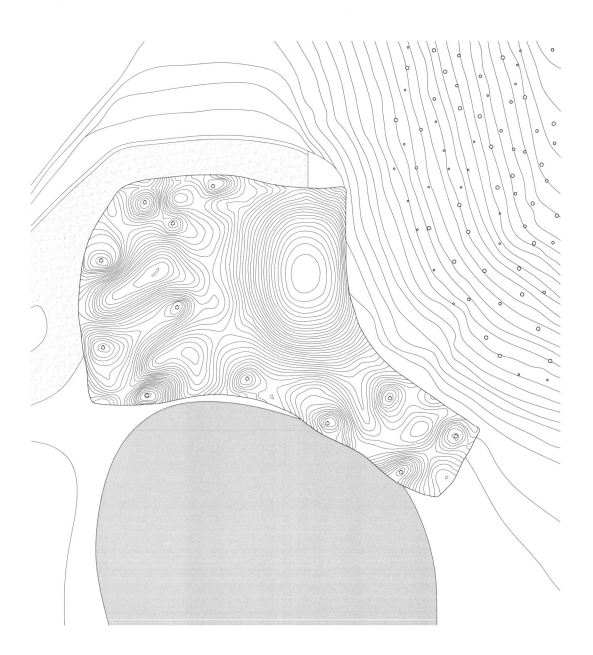

Thin Shell Concrete / Case Study

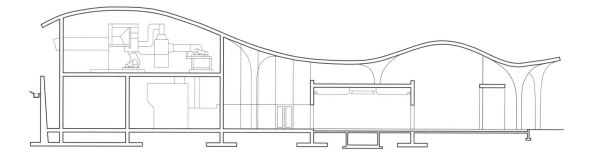

architect gives a significant amount of design responsibility to the computer. Yet no matter what ideological consequences this may imply for the architect's creative authorship, it is a rigorous use of technology in which the role of digital tools is conceptually integral to the formal creation of the final construction.

IN BETWEEN LAKE AND HILL, IN BETWEEN GROUND AND SKY

The interaction between program, architecture and landscape formulates the poetic concept of this structure. In plan, the irregular column locations look like a grove of trees growing naturally in random places, continuing the adjacent forest into the building. Its space becomes an architectural formulation of the place in which the wooded hill meets the lake. The lakeside glazing follows the curved edge of the water, while the retaining wall that faces the hill traces its curved

2.20 The building is an architectural formulation of the place in which a wooded hill meets an artificial lake.

2.21 The smoke from the furnaces gets discharged laterally towards the hill through pipes in the back wall. It blurs the boundary between the artificial topography of the concrete shell and the natural topography of the hill.

topography. Both sides of the building curve outwards and are thus defined through the adjacent landscape features, rather than through the interior organization of the program. The spaces solidify gradually from ethereal and fluid at the lake to solid and dense at the mountain. Above them the roof projects outwards beyond the walls and seems to be autonomous from the inhabitable spaces of the program it shelters. The smoke from the furnaces gets discharged laterally towards the hill through pipes in the back wall. It rises gradually behind a roof overhang, blurring the boundary between the artificial topography of the concrete roof and the natural topography of the hill.

Just as the plan shows the space of the funerary hall as an in-between of lake and hill, the elevation renders the same space as the in-between of ground and sky. The straight horizontal line of the floor distills the cemetery ground-plane, with the valedictory standing on it as a gravestone in architectural scale. In contrast, the curved concrete roof is shown as the lowest hill of the valley beyond. The single lines of the hills in the drawing are meant as an abstract limit rather than a physical condition; they represent different stages of a gradual dissolution

from ground to sky complementing the solidification from lake to hill in plan. The rooms for preparing the cremation are shown as the upper limit of the ground, while the roof from which the smoke of the cremation rises is the lowest downward limit of the sky.

The distillation of the actual ground of the earth into the architectural ground of the floor occurs as a process that the visitor experiences when approaching the building. A curved walkway that leads to the entry has a quasi-natural flagstone paving, which detaches itself from the ground as a clearly articulated slab as it glides under the roof. The rough stones of the paving transition here into a polished marble cover.

The independent column locations both on the building floor and next to it emphasize that the slab is not only a plinth to serve the roof above, but also an independent architectural element that is as meaningful as the roof.

SMOKE AND CONCRETE

The charged space of the crematorium represents the in-between of ground and sky as an architectural compression of the entire world in a threshold to the beyond. The rites performed in this space honor a departed person crossing the threshold by going up in smoke. A traditional cremation hall might be housed in a solid building typically featuring a symmetrical shape that

represents the emotional heaviness of the program, with a central chimney from which the smoke of the furnaces rises. Ito does the opposite: he creates an architecture that is extremely light and seems to consist of the smoke itself. The curved white concrete roof represents the smoke, and the other components of the building each have a distinguished role in relation to this metaphor.

Only the cremation room in which the dead bodies are incinerated connects directly to the roof shell. Constructed

2.22 The programmatic components of the Meiso No Mori Municipal Funeral Hall are designed as individual building volumes that stand independently under an undulating concrete shell. This roof seems to consist of smoke gliding across the site.

out of the same white concrete, its walls unfold seamlessly at the top into the curved roof. Within the same line of movement the roof melts back down and transitions into a column. This configuration is an architectural interpretation of the cremation process: first the cubic *form* of the cremation room dissolves itself in the *formless* shape of the roof; then the roof condenses and falls back down as the circular *form* of the slim column, just as moist air condenses and rains down from a cloud.

The respective programs of the rectangular room and the circular column affirm this interpretation: the cremation that occurs in the room is the physical dissolution of the human body, which

goes up into the air. When it condenses, this air falls back down on the roof as rain; and the rainwater drains in downspouts that are embedded in the columns. The columns do not only artistically represent the condensation process of a cloud hovering above; they literally collect the rain from the surface of the roof and carry it down. Both conceptually and functionally, the columns are an embodiment of the condensation. The drainage pipe inside the column sits within a steel tube that acts as a structural core for the vertical support. This steel core is crucial to achieving the thin column shape, which gives the roof its sense of an anti-gravitational sheet.

A concrete strip footing that runs along the edge of the floor slab features a thin cantilever. This subtle projection creates the perceived edge of the floor from the outside, and produces a shadow reveal that visually detaches the slab from the ground. The straight edge of the floor slab complements the curved edge of the roof shell, since both members have the same thickness and are detached from the ground. This design feature is important for the experience of the extravagant roof shape. It makes the floor appear as an element that could potentially also be curved, and the roof as an element that could potentially also be straight, a duality that enforces the dynamic appearance and sculptural character of the structure. By making floor and roof look identical, Ito removes common associations with what they represent as building components and makes them appear as phenomena of an abstract landscape.

Thin Shell Concrete / Case Study

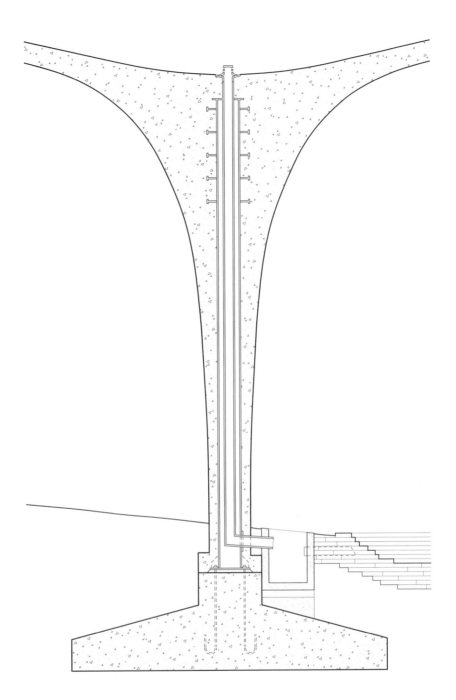

2.23 The walls of the cremation room connect seamlessly at the top into the curvature of the concrete shell. Within the same line of movement, the roof transitions down into a column.

2.24 The columns collect the load of the concrete shell and the rainwater from its surface. A drainage pipe sits within a steel tube that acts as a structural core for the vertical support.

Meiso No Mori Municipal Funeral Hall

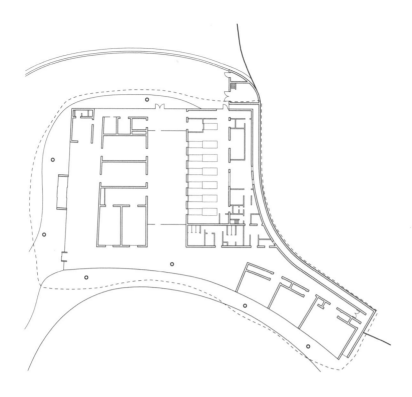

EARTH AND MARBLE

In their incredibly sculptural shape, the white concrete roof and columns do not have any features that give the observer a clue to the scale of the structure. The marble covered floor plate and the volumes that grow out of it are far less abstract: Prominent door openings in the valedictory wall reference the human scale, making this part of the building look like a house standing under a white cloud of smoke that is hovering above it. This house-like appearance enhances the sculptural reading of the roof and the glass, which are perceived as built incarnations of the natural environment. Just as the cremation chamber has a straight edge at the floor and transitions seamlessly with a curve into the white concrete roof that it carries, the marble-clad valedictory has a straight limit at the top and merges with the marble floor that it stands on. Instead of being skybound like the concrete, the marble seems to be earthbound. This gives physical presence to the program, since the valedictory is the last place where the dead body is still physically present. The construction details of floor slab and valedictory are an artistic abstraction of the coming from the earth, just as the cremating room with the roof shell is built as an artistic abstraction of the going into the sky.

WATER AND GLASS

The juxtaposition of building elements and natural elements enhance each other throughout the building. The comparison of the roof to clouds or smoke reads clearly in combination with the reflection of the water in the glass. Water and glass are both liquids; the transparency and

shiny surface character that is common to both render this affinity visually. In order to support the expressive quality of the concrete roof shape, the climatic boundary between indoors and outdoors is designed to appear as ephemeral as possible. Along the lakeside, a glass wall cuts a curve through the space under the roof. It defies the appearance of a common glazing enclosure and presents itself as a pure surface. Ito makes a great effort to conceal the framing profiles for the nineteen millimeters thick toughened glass. At the bottom the glass panes attach to

2.25 In the plan, the glazing follows the curved edge of the water, while the retaining wall that faces the hill traces its curved topography.
2.26 The straight edge of the floor slab complements the curved edge of the roof shell, since both members have the same thickness and are detached from the ground.

the outside of the floor slab, with a steel connection that is anchored to the face of the concrete cantilever. This connection is covered with steel sheets painted white to match the edge of the white concrete shell roof. At the top, the glass panes slide into a pocket between two stainless steel sheets that run along the curved underside of the roof. An additional layer of leveling mortar is applied to the ceiling on either side of the steel sheets to cover the framing. The combination of straight floor and curved roof creates glass panes of varying heights. Under horizontal wind loads each of these sheets has a different bending radius and wants to deflect in proportion to its dimensions. Since there is no strong framing along the vertical edges of the panes, high stresses would occur in the joints between them. A series of metal fins cantilever downwards from the roof at the joint lines to solve this problem. They give the glass a continuous horizontal line of lateral support at the top and make the differently sized panes span from the floor up to the same height. In the context of the landscape set up by the sky roof and the earth floor, the fins appear as raindrops running down the surface of the glass.

COLUMN FORM

As much as the built elements of the funerary hall celebrate the elements of the world as architectural form, the construction process is a celebration of their creation through a series of artistic acts. The building exemplifies remarkable achievements of contemporary thin shell construction not only in the way its shape is designed with the support of digital tools, but also in the way this shape is executed with inventive formwork technologies and chemical concrete admixtures. Reinforced concrete is an ideal technology to build such a freely designed shape, since the material itself imposes no formal restrictions. The real challenge lies in creating formwork on which to pour the complicated shape. The irregular curvature of the funerary hall does not conform to any systematic geometry; thus the forms for the concrete shell cannot be assembled from straight members. Repetitive use of a standardized formwork segments is also impossible, since the curved pieces have to be different for each location in the irregular shape.

Although the entire concrete structure is built as a continuous member from the bottom upwards, different formwork

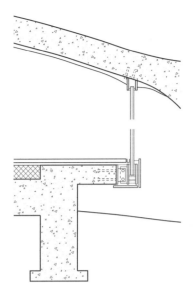

types are used for the individual columns and for the roof shell. The column forms are fabricated in the shop, while the roof formwork that attaches to them is constructed directly on site. In both cases, complex substructures and installations are necessary to hold the mold in its correct position and bend its surface into the required curvature.

The form for one column consists of three vertical portions that form rings around the column shaft; each of these rings is assembled from shop-produced plywood segments. For the proper installation on site it is important that the prefabricated formwork segments are small enough to be handled easily by the construction workers; the bigger the pieces are, the harder it becomes to connect them with a high degree of precision. The four top segments of the column form exemplify how many plywood pieces are necessary to generate the required curvature. In order to produce a tight radius, thin plywood has to be

2.27 The metal framing that connects the glazing to the concrete is concealed to make the boundary between indoors and outdoors appear as ephemeral as possible.

2.28 The assembled column form appears as a vessel that stands in the landscape and captures the elements. It anticipates the drainage function of the concrete column that will be poured into it.

Meiso No Mori Municipal Funeral Hall

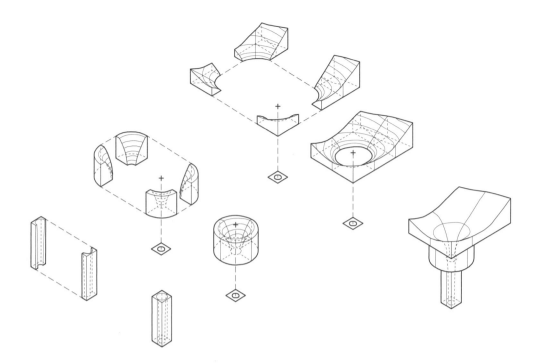

forced into shape with a strong corset that consists of vertical planes pointing to the center of the column. The upper edges of these individual sheets are cut in curved profiles; in combination all the single curvatures define the double curvature of the form surface. For the assembly of the corset pieces a horizontal sheet acts as a base and a curved plywood strip connects them at the top. In their combination these plywood pieces exhibit a tectonic logic similar to the overall configuration of the project: a flat bottom surface combined with a curved top surface, just like the floor and roof of the building.

The assembled column form is an architectural project in itself. As a vessel that stands in the landscape and captures the elements, it anticipates the drainage function of the concrete column that will be poured into it. All the programmatic meaning of the architecture is already fully embodied in it. At the same time the form refers to historical column prototypes of antiquity; it exhibits a traditional separation of the *stylos* (column shaft) and *capital* (crowning member). In this respect the formwork contrasts with the concrete column that is eventually poured into it, which shows a seamless transition to the roof it carries.

DEATH MASK

Ito depends on sophisticated carpenters to produce such a complicated formwork. He uses traditional Japanese wood architecture as a form onto which he pours his contemporary Japanese concrete architecture. The funeral hall is thus a concrete death mask of traditional Japanese woodwork. The product of highly skilled carpentry finds its final resolution in a durable imprint of artificial stone. In this sense the formwork is the actual

architecture, and the concrete is only there to document it in order to permanently preserve its memory, just as a death mask is made to preserve the facial features of
a departed person.

A continuous striation of planks makes up the roof formwork. These planks connect seamlessly to the column form. They are held in place by a substructure of curved ribs that look like the lines Ito draws into space on his elevations. Each rib consists of two plywood strips with the same curvature in different segments that are lapped in layers and tied together to create a continuous member. Shoring posts standing on a

2.29 The form for one column consists of three vertical portions that form rings around the column shaft.
2.30 Each ring is assembled from shop-produced plywood segments. They consist of many pieces to generate the required curvature.

temporary floor lift the ribs to the right height and wooden rafters stabilize them laterally. The final formwork layer of thin planks is bent over this substructure. Even within a small segment of the roof surface each piece of the form has a double curvature; nothing is facetted out of straight members. The finished roof formwork is a topography of artificial timber hills and valleys overlooking the similarly curved terrain of the surrounding landscape.

COLUMN AND ROOF REINFORCEMENT
The steel reinforcement of the column makes a funnel-shaped version of a *tied column*, in which circumferential reinforcing rings tie the vertical bars to their required positions to prevent them from buckling under load. The steel tube that is placed in the center of the column as a structural core features lateral metal pins for proper bondage to the concrete.

Thin Shell Concrete / Case Study

Radial reinforcing bars across the top of the funnel in the formwork connect the steel tube to the outer reinforcement rings to hold it in the right place while the concrete is poured. The drainage pipe is placed as a separate tube sitting within the structural member, so that it does not have to bond with the concrete. If the drainage pipe leaks it can be pulled out from above and replaced.

For continuity in the concrete structure, the vertical bars of the tied column connect to the reinforcement of the roof shell. Because of its relatively flat curves and asymmetrical shape, this concrete

2.31, 2.32 Ito uses traditional Japanese wood architecture as a form onto which he pours his contemporary Japanese concrete architecture. Top: assembled column form. Bottom; column and roof formwork.
2.33, 2.34 The hills and valleys of the roof formwork consist of wooden planks held in place by curved ribs.

shell performs partially in bending and creates high stresses in the material, especially where it meets the supports below; this means that stresses other than purely membrane stresses apply. For this reason the shell is almost three inches thick and has a double layer of reinforcing bars embedded into it to resist flexural stresses. Additional bars support the areas in which especially high stresses occur. Their direction and location at the top, bottom, or both at top and bottom of the shell section depends on the stress distribution within the member. As a pixilated landscape of forces, the reinforcing plans render the digital creation of the shell curvature in its "low resolution." The formwork is used to shape the reinforcing and hold it in its correct position: *bar chairs* tie the steel to the wooden planks in regular increments and translate the curvature of the plywood into the metal.

RAPID-HARDENING CONCRETE

As is typical for shell constructions, the concrete is placed directly on the single bottom formwork. It needs to have a very precise mixture, liquid enough to distribute evenly in the tight cavities of the column forms, but also hardening quickly so as not to run down on the sloped surfaces of the roof forms. For this purpose, chemical *accelerators* are added to the concrete; they quicken the hydration process in which the concrete hardens and give it a *high early strength*. A concrete with this kind of *admixture* must be placed with accurate timing. The whole structure is poured in five upwards increments starting at the columns. Upon completion of the pours, thin layers of insulating mortar and a polyurethane coating are applied to the top surface of the concrete shell. These layers compensate for two significant deficiencies of

the roof, giving the concrete better thermal insulation properties and making the porous material waterproof. The partially removed formwork on the underside of the roof constitutes again a state of the project that is almost too beautiful to be temporary. The entirely monolithic concrete shape that appears upon the removal of all the remaining plywood looses the directional texture of the wooden planks that was so powerful in its dynamic expression of the curved surface. With the spackling of any remaining texture marks, the erasure of formwork traces from the concrete surface is complete. Although this loss is sad, one has to consider the funeral hall as a continuously evolving art piece that has several incarnations during its different phases of construction. Each of these incarnations is powerful but has to vanish in order for the next one to carry on

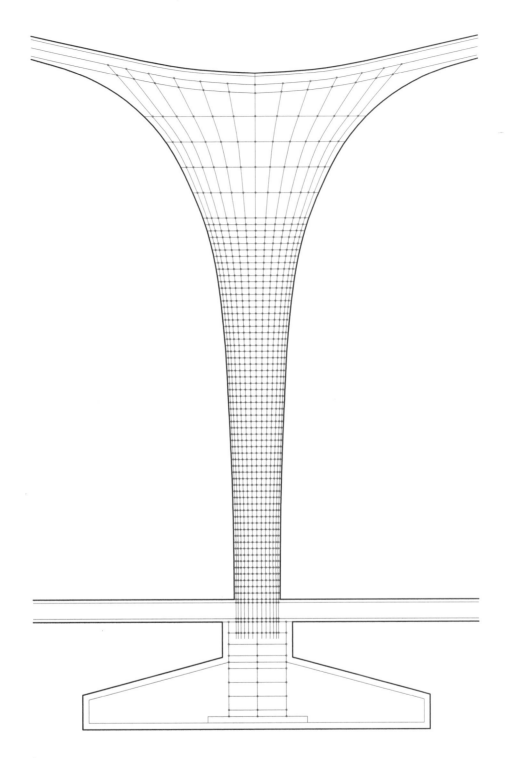

2.35 Circumferential reinforcing rings tie the vertical bars to their required positions to prevent them from buckling under load.

2.36 For continuity in the force flow, the reinforcing of the tied column connects to the reinforcement of the roof shell.

Meiso No Mori Municipal Funeral Hall

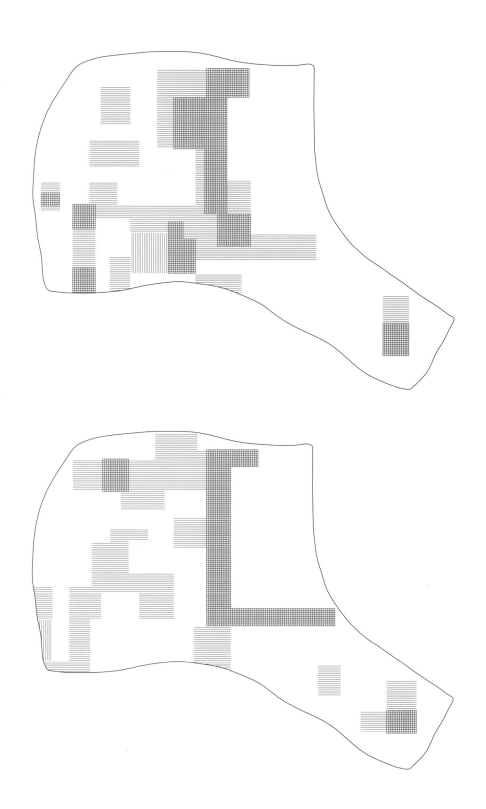

Thin Shell Concrete / Case Study

2.39

2.38

2.40

2.37 Location and direction of additional steel reinforcing at the top (top) and bottom (bottom) of the concrete shell in areas of high stresses. As a pixilated landscape of forces, the reinforcing plans render the digital creation of the shell curvature in its "low resolution."

2.38 Chemical admixtures make the concrete harden quickly so as to not run down on the sloped formwork surfaces.

2.39, 2.40 Through the removal of the formwork, the concrete shell loses the memory of its construction right after it has been completed.

The formwork for the retaining wall rises against the hill; the slab on grade and the strip footing along its edge have been poured; the formwork under its edge cantilever is still in place; the reinforcement for the walls goes up; the steel cores of the columns stand on their individual footings.

The linear ribs for the roof formwork are constructed on a scaffolding platform; gaps are left in the platform for the shop-fabricated column forms; the reinforcing for the back wall of the furnace room rises beyond.

The formwork for the furnace room walls is erected simultaneously with the planking of the roof formwork.

A truck pumps concrete up to the desired level. Work starts with the columns and low parts of the shell and goes up to higher elevations in five pours.

The structural concrete work is completed.

The finish layers are applied to the roof surface, which becomes smooth and bright.

All the finish work on the concrete is complete and the scaffolding can be removed; the glazing is installed and the work on the surrounding landscaping can begin.

It is a miracle that this project got built...

the idea in a different way. The memory that remains from each phase to the next is not a literal trace of the previous condition, but rather a new formulation of the continuously present artistic identity of this building.

TIMELESS

A concrete shell structure immediately loses the memory of its construction right after it has been completed. The actual construction work of the form-work does not remain as a permanent component of its architectural matter, but is removed and discarded when the concrete has fully hardened.

The physically present material of the thin concrete shell does not incorporate any elements that refer to the builders that have put them in their place. The human effort that created this architecture is not immediately comprehensible. With the absence of a tangible construction history, the shell gives no indication of the time necessary to bring it into the world, as a masonry wall or timber frame does. A lightweight roof built as such a concrete structure creates an environment that is dematerialized through its minimal physical presence, and de-historicized in its negation of time. This *time-less* character forces the inhabitant

to search for an understanding of the architecture as a found landscape, rather than a built creation. We cannot put ourselves in the role of a hypothetical builder of its curved spaces, but have to instead accept it as a part of the world that is given to us, just like a part of the natural environment.

2.42 The absence of a tangible construction history gives thin shell concrete structures a time-less appearance.

Thin Shell Concrete / Case Study

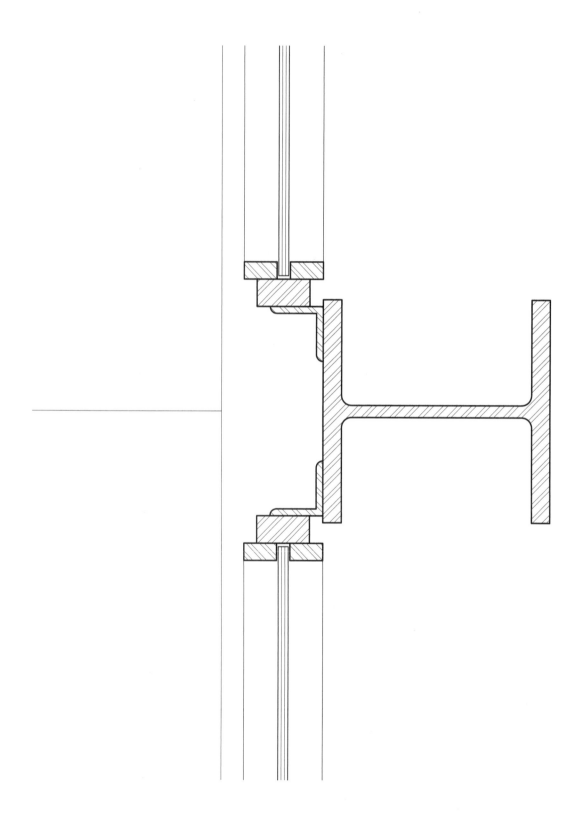

FARNSWORTH HOUSE
GALLERY OF THE TWENTIETH CENTURY

SENDAI MEDIATEQUE

STEEL FRAMING

The strongest construction materials
come from the ground. A masonry stone
is cut from rock. Brick is molded from
clay. Iron is smelted from an ore that is
found in the Earth's crust. This creates
a powerful metal that has been used in
construction since Antiquity. Blacksmiths
forged wrought iron into fasteners that
held together the stones of ancient
temples and drew it into chains that
stabilized the domes of the Renaissance.
The furnaces of the Industrial Revolution
made iron easy to cast and strong in
compression. This new cast iron was
shaped into columns and arches that
made up the great bridges and lofty
vaults of the Machine Age. With the
advent of modernity, the precise adjustment of the carbon content in the iron
alloy created an even more powerful
metal—structural steel. Steel combines
the strength of cast iron with the toughness of wrought iron and can be used
for virtually any kind of building member.
The strength of steel maximizes the scale
of architecture, while minimizing its
physical presence. Steel is a distillation
of the hardest substances of the earth.
The structures built from it condense
architectural form into a thin spatial
framework that is both conceptually and
literally the essence of modern building.

Steel Framing

ROLLED STEEL SECTIONS

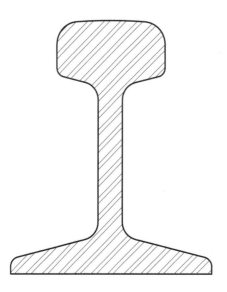

Before the Renaissance, steel tools and weapons were produced with various techniques in Africa, Europe and Asia. In the seventeenth century more efficient production methods increased the use of steel for a broader range of purposes. It was not until the nineteenth century that steel became a mass-produced material. The powerful blasting furnaces of the *Bessemer process*, in which air is blown through molten iron to remove impurities, increased the efficiency of steel production. It could then be used for many building components. The costly production and enormous weight of steel demanded efficient shaping to reduce the required amount of material. For this reason steel is typically linear, getting its character from a section profile. These sections anticipate a particular use in structural framework; they are designed to resist a maximum force as a beam

or column with a minimum amount of material. A configuration of structurally active planes called *flanges* are held together in a certain offset by a *web* in order to create a large section profile. The geometric logic of each section type remains consistent throughout all scales.

Steel is shaped into these section profiles in various ways. A steel plate can be cut into strips that are assembled into a *composite section*. This technique creates steel members of virtually any size and shape. It is ideal to custom-produce special elements that have complex configurations of webs and flanges. But the several stages of this assembly are not ideal for industrial production. *Structural shape rolling* is a more efficient method for mass-producing steel sections. A rectangular metal piece is passed through pairs of rolls for the desired shape in the foundry, while it is still warm and

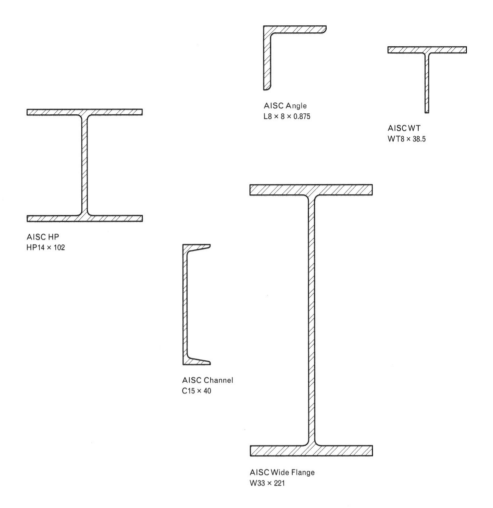

AISC Angle
L8 × 8 × 0.875

AISC WT
WT8 × 38.5

AISC HP
HP14 × 102

AISC Channel
C15 × 40

AISC Wide Flange
W33 × 221

3.1 (previous page) Steel structures condense
architectural form into a thin spatial framework.
3.2 Rail-steel is the ancestor of the rolled I-beam:
it has a base to stand, the equivalent of the lower
flange, and a head for the wheel to roll on, equivalent
of the upper flange.
3.3 Section profiles such as L's (angles), T's, C's
(channels) and I's are designed for a particular use
in a structural framework.

3.4

malleable from the casting operation. Pre-made dies are changed at several stages of a gradual process in which the metal passes backward and forward through the mill and deforms into the desired shape. The result is a long steel member with a consistent section-profile. While the metal is still hot it is cut to a length that can be handled. Then it is cooled, straightened, and cut to the ordered length. In modern foundries all stages of this production occur in one continuous sequence, from the iron ore to the finished steel member.

This procedure originates from *rail steel*, the oldest rolled section profile. It is the ancestor of the *I-beam*, since its shape suggested its use as a beam that resists bending stress: It has a base to stand, which would be the equivalent of the lower flange, and a head for the wheel to roll on, comparable with the upper flange. The distinct top and bottom are connected with a stem, alike the web. Engineered for railroad tracks, rail steel provided an infinite steel extrusion that could be cut to any length. In the 1850's the American inventor Peter Cooper used rail steel for the beams in his buildings and further developed this section type for construction use, which created the I-beam. Soon a variation of this shape with wider flanges was available for use as columns, and the entire structural frame of a building could be composed from steel.

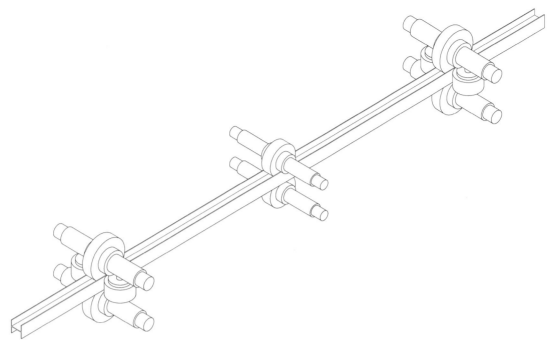

3.5

3.6

3.4, 3.5 In structural shape rolling the warm steel is
passed through a series of rolls that have the reverse
of the desired shape.

3.6 Over several passes through a rolling mill the
section of a rectangular work-piece transforms gradually into a structural shape.

Rolled Steel Sections

FARNSWORTH HOUSE
GALLERY OF THE TWENTIETH CENTURY

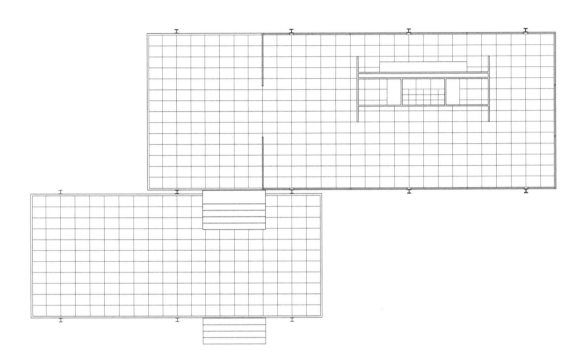

The Farnsworth House in Plano and the Gallery of the Twentieth Century in Berlin—now called New National Gallery—are explorations of the artistic potential of structural steel sections. The two Ludwig Mies van der Rohe buildings share many principles in design and construction. Their exposed steel columns and beams create a bold expression of the forces that flow through their structural system. As modern incarnations of archetypal concepts, they distinguish clearly between vertical members that carry, and horizontal members that span between them. The strength of steel allows Mies to distill architectural form to an essence that is beyond time and style. He compresses mass and meaning in the overall composition of the buildings, as well as in their individual members. Despite this common approach to the strength of the metal, house and museum embody different techniques of steel construction. The contradistinct ways in which they are built and proportioned celebrate their respective rural and urban environment.

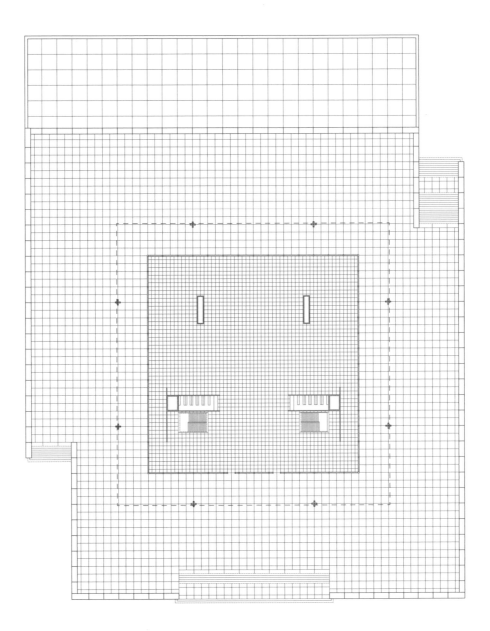

3.7 Farnsworth House, Plano IL, 1949
Rectangular
Groundless
Directional
One-way flexural system
Rigid column-and-beam frame connections
Rolled steel sections
Cut to shape and length
Straight members
Ionic
White

3.8 Gallery of the Twentieth Century, Berlin, 1969
Square
Grounded
Non-directional
Two-way flexural system
Rigid plate on pinned column
Composite steel sections
Assembled from plates
Curved and tapered members
Doric
Black

The summerhouse of Edith Farnsworth stands at a river shore in a picturesque landscape. Its rolled steel members are assembled to form an architecture that is an artistic celebration of the site. The structure acts like the frame of a landscape painting that captures a piece of nature as a tangible subject. Architecture becomes a bodiless *frame-work* that registers the environment but does not establish any spatial boundaries. The vertical planes between the columns and girders are transparent throughout and only partially filled with glazing. Within the framing of the platforms the horizontal surfaces are paved with limestone, creating an inhabitable ground in the uninhabitable terrain of the river's flood plain. The site plan of the house does not show the edge of the river or any other features of the ground. The house stands virtually in the river and the only "ground" in the plan is the

steel platforms themselves. Parallel to the river, they shift in a dynamic movement, as if they were not fixed to their location; the smaller platform appears to be floating away with the current of the water.

The surrounding landscape provokes the rectangular proportions and structural grid of the Farnsworth House: three rows of columns run parallel to the river. They are connected with girders, which support smaller cross beams. Precast concrete planks create the surface of the platforms. All elements function either as columns or beams in the hierarchy of a *one-way flexural system*. This type of framework lends itself to the construction of a rectangular building: large bending members span between vertical supports in the main direction, smaller bending members span between them in the opposite direction. Not only does this configuration establish a directional logic that relates to the river. It also

Steel Framing / Case Studies

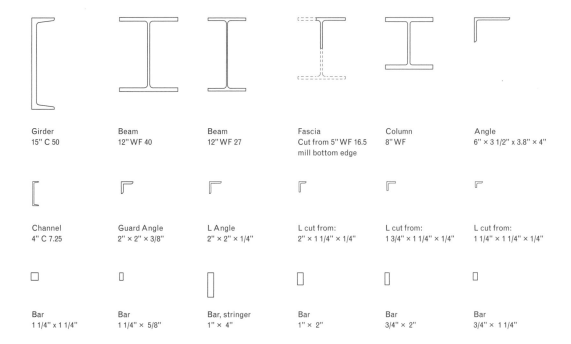

Girder 15" C 50	Beam 12" WF 40	Beam 12" WF 27	Fascia Cut from 5" WF 16.5 mill bottom edge	Column 8" WF	Angle 6" × 3 1/2" x 3.8" × 4"
Channel 4" C 7.25	Guard Angle 2" × 2" × 3/8"	L Angle 2" × 2" × 1/4"	L cut from: 2" × 1 1/4" × 1/4"	L cut from: 1 3/4" × 1 1/4" × 1/4"	L cut from: 1 1/4" × 1 1/4" × 1/4"
Bar 1 1/4" x 1 1/4"	Bar 1 1/4" × 5/8"	Bar, stringer 1" × 4"	Bar 1" × 2"	Bar 3/4" × 2"	Bar 3/4" × 1 1/4"

reflects the linear character of the rolled steel section as such. All elements of the framework have a determined structural function and can be specified as an industrially rolled profile. The house becomes a catalogue of Is, Cs, Ls and rectangles that are used for different parts of the system. Their shape exhibits their structural function and their size reflects the magnitude of the load they have to resist.

3.9 The Farnsworth House is a bodiless frame-work that registers the environment without establishing any spatial boundaries.

3.10 The steel assembly of the house is a built catalogue of rolled Is, Cs, Ls and rectangles. The shape of each section exhibits its structural function; the size of the section reflects the load it resists.

3.11 The mirror image of the Farnsworth House in the river "reflects" the structural identity of its framework; it can be turned upside down like a steel I-beam.

3.11

Seen from across the river, the steel members of the house also speak to the features of the landscape: slender columns repeat the vertical pattern of the surrounding trees and the prominent edges of the platforms correspond to the flatness of the river surface. The trees, reflected in the water, naturally appear turned upside down. But the mirror image of the house remains unchanged. This literally "reflects" the identity of its steel structure. When turned around, it would look and perform in the same way. The house interprets the structural concept of a steel beam at architectural scale: the platforms resemble the flanges and the columns resemble the web. A concrete beam needs to be reinforced where it is stressed in tension, either in the upper or lower part of its section, depending on the load condition for which it is designed. In a steel I-section or C-section the upper and lower flange have the capacity to handle both tension and compression and are thus interchangeable. As its enlarged rendition, the roof and floor of the house are interchangeable as well, creating architecture that can be turned upside down.

The architecture of the New National Gallery is grounded in the city. A base story that contains the museum's permanent collection is partially submerged in the terrain of the site. The sloping ground condition forces Mies to engage the site in a way that is in contrast to the groundlessness of the place in which the Farnsworth House stands. A granite-

clad base creates a flat platform in the irregular topography. Its heavy appearance complements the lightness of a flat steel roof that hovers above it, suspended by a magical force. This architecture cannot be turned upside down. It represents a traditional relation of architectural form to gravity: where the building meets the ground its elements become solid and heavy.

The steel roof spans two hundred and twelve feet; yet it is supported on only eight columns. This provokes a surreal impression. Its horizontal expanse attempts to shelter the entire city, rather than only being the upper limit of a single building. It covers a glazed hall for temporary exhibits, in which the artwork of the twentieth century is exhibited as an integral part of the continuous streetscape, visible for every passerby. This constitutes a spatial center for the fragmented urban surroundings of the museum. The exhibition hall contextualizes the city around it, turning it into the backdrop for the artwork that is exhibited under its roof.

The columns and roof girders of the New National Gallery are not arranged in a directional logic such as that of the Farnsworth House. A square plan

3.12 The roof of the New National Gallery is a thin steel plate supported on eight steel columns; it hovers over the city suspended by a magical force. A masonry clad base story that contains the museum's permanent collection is submerged in the terrain. It creates a flat platform for the steel structure that houses a hall for temporary exhibitions.

organizes all components in a neutral orientation. Two columns stand in the same intervals on each side of the square roof. Such a framework has no logic for spanning large girders in a major direction that support smaller beams in the other minor direction. A series of equally strong girders intersect in a grid pattern that distributes the load to the columns on each of its four sides; as a *two-way flexural system* the roof spans in both axes simultaneously.

Standard rolled sections are produced to resist only one kind of stress. A *two-way flexural system* lends itself to be composed from custom sections that are specifically designed for the varying stresses in each segment of the structure. Both the roof and columns of the New

National Gallery are therefore composite steel members that are assembled from individual steel plates, rather than specified from industrially rolled section profiles. While the structure of the Farnsworth House is built from industrial sections that are selected for the use as an element of structure, the columns and girders of the New National Gallery are in themselves already "built" as an integral part of the construction sequence.

STEEL COLUMNS AND SLABS

The compressive strength of steel allows for slender columns that resist large vertical loads. Both in the Farnsworth House and in the New National Gallery a thin stick would be sufficient to resist the weight imposed on them. But such

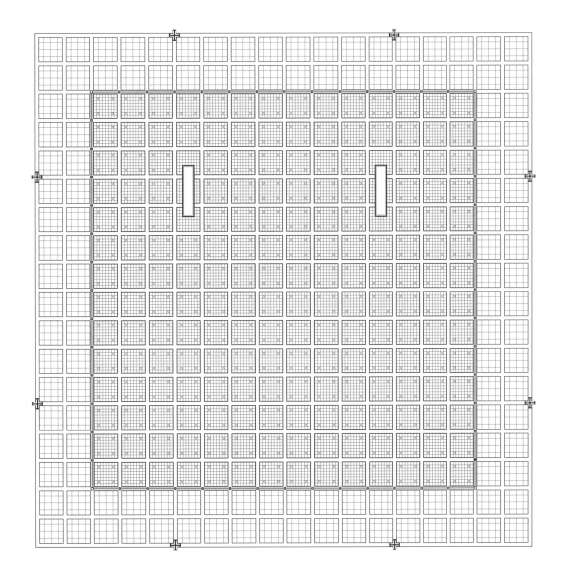

3.13 The roof of the New National Gallery consists of two sets of seventeen girders that intersect in a two-way flexural system.

3.14 The grid of the roof structure is continuous throughout the interior and sheltered exterior of the exhibition hall.

a column would buckle and fail, when it is pushed to the side by a horizontal force. Since neither building has any internal walls to stiffen their structure horizontally, the columns have to resist all wind loads pressing against the glass envelopes. This stiffening function determines their distinguished shape. Flanges and webs act as small walls that have been evacuated from the interior of the building and compressed to the scale of a construction detail.

In the rectangular Farnsworth House the long elevations catch more wind than the short elevations. A rolled I-section provides an ideal shape for a stiff column in such a configuration. When it stands upright, its strong axis resists horizontal forces in one direction, just as it would resist vertical loads when used as a beam. The square New National Gallery has equal elevations on its four sides and

accordingly the wind loads are identical in both directions. If the columns were made from the same type of industrial profiles as in the Farnsworth House, this would require two I-sections with opposite orientations, one for each axis of the square structure. That is what Mies designs: a custom-made cruciform section, resembling two I-sections that intersect in their centerline. It reads as a transformed version of the masonry base: the steel planes extrude the intersecting grid lines of the stone pavers up to the roof.

Instead of formulating a bold contrast between a heavy base and a suspended roof, all horizontal planes of the Farnsworth House are constructed as the same thin steel slab. Floor and roof have to be identical in an architecture that can be turned upside down. The lowest slab is an open terrace,

elevating the inhabitant off the ground. A second slab on a higher elevation creates the platform of the house's porch and interior. A third slab of the exact same size acts as a roof above it. Each slab is framed with the same rolled C-section as an *edge-girder*. This steel member is fully exposed to emphasize its role as an armature for the slab's surface. Because of its function as a bending member it is larger than the I-section used for the columns. A girder is stressed mainly at the top and bottom

of its section, rather than across its entire surface.

Although the three slabs are constructed from the same components, subtle variations at their edges distinguish between their different architectural functions. On the low terrace, the travertine pavers are raised half an inch above the top of the edge girder. This emphasizes that the primary purpose of this platform is to lift the inhabitant off the ground. On the upper platform the pavers are flush with the top of the channel. Another element projects into the space of the porch instead: the ceiling is half an inch down from the bottom of the roof edge girder. This creates the same condition that exists in the top of the lower platform turned upside down. It emphasizes the shelter from above that the roof slab provides for porch and interior. On top of the roof edge

3.15 An I-section has an ideal shape for a stiff column in the rectangular Farnsworth House. When it stands upright, its strong axis resists horizontal forces in one direction.

3.16 The cruciform column of the square New National Gallery resembles two I-sections that intersect in their centerline, one for each axis of the building. The steel planes extrude the intersecting grid lines of the stone base up to the steel roof.

Steel Framing / Case Studies

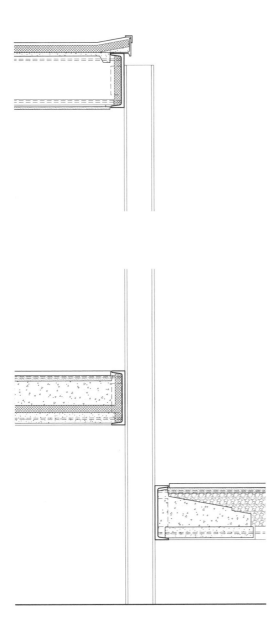

an additional set of steel sections captures the waterproofing. This capping projects beyond the edge of the frame. It articulates the function of this slab's upper surface as a shield from rain and sunshine, rather than as an inhabitable plane contained within the frame.

In the slabs of the Farnsworth House only the edge girder is visible and the smaller cross beams are covered. The steel structure of the New National Gallery roof is fully exposed. An expressive structural grid runs continuously throughout the underside of the roof, both within the exhibition hall and outside in the roofed exterior. Two sets of seventeen girders intersect in a grid of *coffers*. In order to make the entire roof span as a single structural unit, the girders connect rigidly at their intersections. They would deform in *torsion* were they not held in position laterally. A continuous steel *compression plate* connects all girders at the top of the roof to prevent them from twisting. The structure is not a crossing set of girders but one large composite steel member, consisting of a gigantic square top flange to which webs and bottom flanges are attached from below. Within each coffer an additional set of smaller stiffening plates attach to the compression plate to prevent it from buckling. This makes the compression plate read as a small rendition of the entire roof system, repeating the overall structural grid in the scale of its construction detail. All components are optimized to make the roof plate as thin as possible. It spans two hundred and twelve feet in both axes with a structural depth of only ten feet.

3.17 The three slabs of the Farnsworth House are framed with the same C-section as an edge girder, although they have different architectural functions. They attach to the sides of the columns as if they could move up and down along them. For consistent elevations Mies wraps the edge girder around the corner, although it does not span between the columns on the short side of the house and therefore does not fulfill the same structural function.

3.18 Subtle variations at the slab edges distinguish between the similar structures of lower platform, upper platform, and roof.

3.19

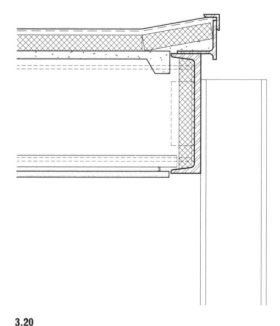

3.20

SUBTRACTION AND ASSEMBLY

The industrial profiles of the Farnsworth House are cut into the required shape in a mode of subtraction. Mies treats the steel profile that was rolled in a mill like a stonemason cuts a marble block that was sliced in a quarry. He respects the already determined proportions of the material and further reduces its primal shape to put its full structural and aesthetic potential into effect. The steel column of the house is derived from a continuous extrusion that is cut to a required length. Its upper end reads as a literal "section" through a continuous member, rather than as a definitive endpoint; the formal character of the I-section becomes tangible architecturally. This detail makes the column appear to be a partial materialization of an infinite line. The imagined continuation of the members in space suggested by this kind of detailing also resonates with the linear architecture of the house as a whole. Its arrangement of steel members could potentially run along the entire river and is only "cut" to the required length.

Some sections are also sliced longitudinally for a customized profile. The fascia piece along the edge of the roof slab demonstrates this method of producing a custom section from a standard section for a particular purpose. It is cut from an I-section that lays on its side, with its web connecting to the edge girder below. One flange of the I-section is cut off entirely, while the other flange is reduced to two thirds of its original dimension. The custom L-section that remains has a drip nose at its bottom that prevents rainwater from running against the face of the edge beam. It also creates a shadow-

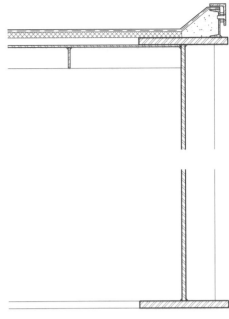

reveal that visually emphasizes the vertical surface of the fascia.

Rather than being subtracted from rolled sections, all members of the New National Gallery are assembled from individual steel pieces in an additive procedure. The cruciform column cannot be rolled industrially and has to be custom-assembled from several plates.

3.19 The upper end of the Farnsworth House column is cut like a "section" through a continuous profile, rather than a definitive endpoint. An invisible plug-weld rigidly connects the flange of the I-section column to the web of the C-section girder.
3.20 The fascia piece on top of the roof girder is a custom L-section that is subtracted from an industrial rolled I-section.
3.21, 3.22 In the steel roof of the New National Gallery a continuous compression steel plate connects all girders. Smaller stiffening plates attach to the compression plate from below to prevent it from buckling within each coffer.

The compression plates, webs and flanges of the roof are individual pieces that are assembled to a single large composite member. The type of steel of these pieces varies to resist different stress levels in the roof while maintaining a consistent appearance of the grid throughout. In some areas the stresses are particularly high, for instance along the edge of the roof where all loads get transferred to the columns. Here the flanges get thicker and are made from steel with a higher *grade*, i.e. hardness. The uniform appearance of the roof that looks like a serial arrangement of identical elements is made possible by a complex non-serial production. Variations in the material compensate for the different stress exposures under the force of gravity. All the horizontal and vertical steel components of the structural grid are displayed along

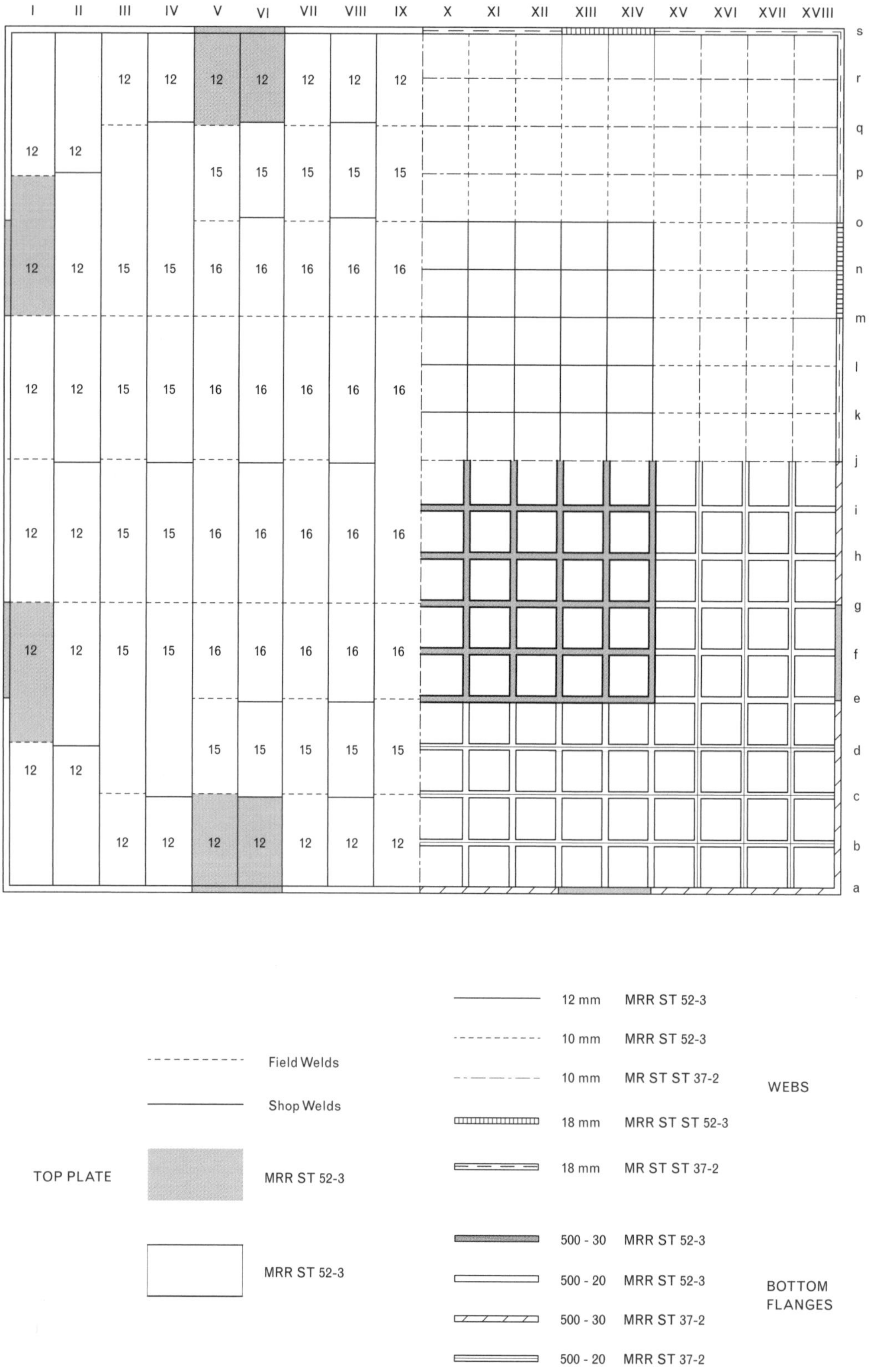

Field Welds

Shop Welds

TOP PLATE

MRR ST 52-3

MRR ST 52-3

12 mm	MRR ST 52-3	
10 mm	MRR ST 52-3	
10 mm	MR ST ST 37-2	
18 mm	MRR ST ST 52-3	WEBS
18 mm	MR ST ST 37-2	
500 - 30	MRR ST 52-3	
500 - 20	MRR ST 52-3	BOTTOM FLANGES
500 - 30	MRR ST 37-2	
500 - 20	MRR ST 37-2	

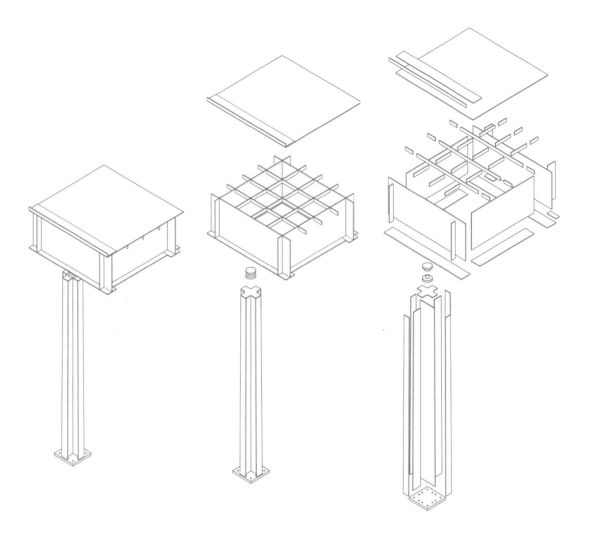

3.23 The grade and thickness of the steel plates that compose the roof varies proportionally to the stress of their location. Over the columns and in the center of the span (areas indicated in grey and black) the steel has to be particularly strong because it resists maximum stresses. The uniform appearance of the roof that looks like a serial arrangement of identical elements is made possible by a complex non-serial production.

3.24 The cruciform column and roof grid of the New National Gallery are assembled from individual steel plates that are continuously welded to composite sections.

the outside of the roof edge. Rather than looking like a definitive frame that contains the roof within its limit, it suggests a furthering of the assembly process to make the roof plane expand in all directions, just as the linear members of the Farnsworth House seem to expand through the way in which they are cut.

WELD JOINTS

3.25

Any steel framework is an arrangement of components that transfer forces in their connections. This makes jointing the most crucial aspect of steel construction. The design of the connections determines which forces are translated and how the structure behaves as a whole. *Rigid connections* transfer forces in all directions and make a framework stiff. In the nineteenth century stiffening plates were *bolted* or *riveted* to the corners of a frame to create this type of connection. The development of *welding* technology revolutionized steel construction in the early twentieth century, because it allowed for rigid connections without any additional plates and fasteners. Welding joins steel members by causing coalescence, i.e. uniting separate pieces into a new whole. The steel elements are locally heated with added filler metal to form a pool of molten material that cools to become a rigid joint.

Different energy sources can be used to create a welding flame, which must be several thousand degrees Celcius to melt the metal. *Gas welding* employs the combustion of acetylene in oxygen; it was

3.26

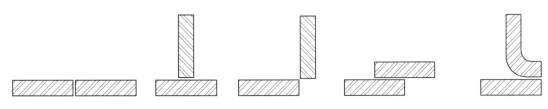

3.27

3.25, 3.26 Arc welding uses a power supply to create
an electric arc between an electrode wire and the
steel to melt it at the welding point.
3.27 The five basic types of weld joints are (from left
to right): butt joint, T-joint, corner joint, lap joint, and
edge joint.

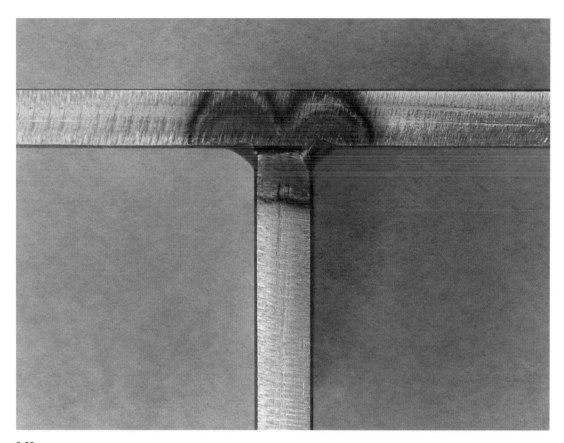

3.28

developed in the late nineteenth century. This technique was popular because of its portability and relatively low cost. As the twentieth century progressed it was largely replaced with *arc welding*, a technology that was substantially advanced during World War II for the production of tanks and submarines. It uses a power supply to create an electric arc between an electrode wire and the steel to melt it at the welding point. In *constant voltage supplies* any fluctuation in the distance between the wire and the base material results in a drastic change in current. If the wire and the base material get too close, the current increases rapidly, which causes the heat to rise and the tip of the wire to melt. In manual construction it can be difficult to hold the electrode steady, and as a result, the arc length and voltage tend to fluctuate. For that reason *constant current supplies* are used on construction sites, because they maintain a relatively stable current even as the voltage varies. In the post war era, metal coverings for the electrode known as *flux*, were developed to stabilize the arc and shield the base material from impurities. This *shielded metal arc welding* is extremely versatile and the most common process in construction.

The five basic types of joints in which steel pieces can be welded together are the *butt joint, T-joint, corner joint, lap*

3.29

joint, and *edge joint*. Each constellation
anticipates a certain placement of the
weld to melt the steel pieces together for
the desired force transfer of the connection. A *fillet weld* unifies steel pieces
along the line in which they meet This
type of welding lends itself to the production of composite steel sections, since
it can merge individual plates perpendicular to each other into the structural
shape of a T, L and other desired configurations. This fusion of several pieces
into a single element also occurs visually;

a weld seam blurs the sharp edges of the
individual elements. Rather than running
along a continuous line, a *plug weld*
connects overlapping plates at certain
points. Holes are drilled into one piece
in the zone of overlap and the welds are
placed through them. The stress in the
connection is distributed over a larger
surface area. The desired welding method
determines the selection of steel elements
for a framework as much as the overall
structural concept. In addition to fulfill-
ing a structural function the shape of the
member has to have the right geometric
properties to connect to the adjacent
members with the desired type of weld.

3.28   In a T-joint two steel plates connect with a fillet
weld.
3.29   In a lapped joint two steel plates connect with
a plug weld.

Steel Framing / Case Studies

RIGID AND PINNED CONNECTIONS

The geometry of a rolled I, T, L or C is always a configuration of intersecting planes. They can connect in lapped joints because a plane of one member can attach to a parallel plane of the adjacent member. This allows them to be plug welded, which has the aesthetic advantage that the individual steel components retain their sharp edges. While Mies uses conventional bolted connections in the

invisible parts of the Farnsworth House, he uses plug welding for all the visible connections. For this purpose the steel profiles are selected and arranged to meet in lapped joints.

In the connection between girder and column, the web of the C-section faces outward so that it can receive the parallel flange of the I-section in a lapped joint. During construction the steel sections are temporarily fastened to each other with *stud bolts* and *cleats*. The flange of the column is perforated with a hole; a plug-weld is placed into it and bonds the two members permanently. As a result an invisible joint rigidly connects the steel elements, although they appear to be only touching each other. The way in

3.30 Rigid connections between columns and girders create the stiff structural framework of the Farnsworth House.
3.31 Mies uses conventional bolted connections for the invisible parts of the framework. All visible connections are plug-welded; this technique determines the selection and arrangement of the steel profiles.

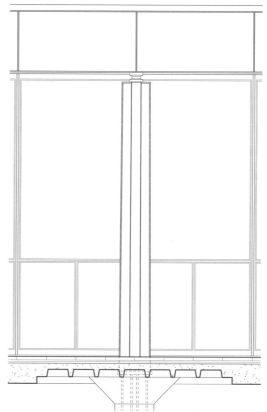

which the slab attaches to the sides of the columns makes it look like a fleet in the water that is only tied to a pole. It suggests a shifting of the house's elements. The top of the columns pass the girder or stop short under its upper limit, suggesting that the slabs can move along the columns, rather than resting on them in a fixed connection. Each horizontal surface of the house becomes an imaginative floodplain that can slide up and down like the river changes its elevation over the seasons.

Rather than implying an infinite line that bypasses the slabs as in the Farnsworth House, the column of the New National Gallery has definitive endpoints at the base and under the roof. At the bottom a rigid connection makes the column cantilever out from below, inducing a higher stress in its bottom part. The shape of the column responds to this higher stress with a slightly larger section at the base. A subtle taper makes the shaft grow slimmer as it rises upwards. While Mies uses welding in the Farnsworth House to connect the rolled sections of columns and girders, he employs it in the New National Gallery to create the structural sections themselves. Individual steel plates are T-jointed to the cruciform section of the column and the structural grid of the roof. The plates are continuously fillet-welded along their seams, generating the composed section that acts as a single piece of steel.

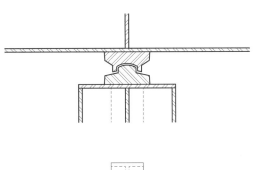

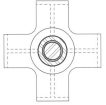

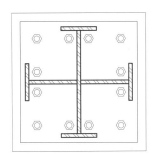

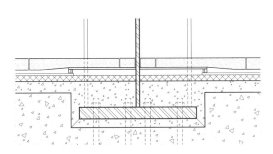

3.32 (opposite) The column of the New National Gallery has a subtle taper that makes the shaft grow slimmer as it rises upwards.

3.33 (above) In its rigid connection to the base, the webs and flanges of the column are welded to a steel plate that is bolted to the concrete structure below. In the pinned connection at the top a cross-shaped head plate holds a circular pin that receives the vertical load and secures the roof laterally.

In its rigid connection to the base, the webs and flanges of the column are welded to a steel plate that is bolted to the concrete structure below. In the pinned connection to the roof a cross-shaped head plate holds a circular pin. This pin receives all vertical loads from the roof and funnels them into the very center of the column, reducing its tendency to buckle as much as possible.

The pinned connection also secures the roof laterally. But unlike a rigid connection it is free to rotate and induces no bending moment. This avoids tensions in the system that could occur if the roof expanded and contracted due to temperature changes. The pinned connection separates the steel structure of the building into two subsystems of column and roof, which are each rigid in themselves but connected with flexibility.

NEGATIVE CAPITAL

In English the word "column" is used for all vertical supports. In the German language the word "stütze" stands for a support with a square or rectangular section, which reflects the origin of such a form as a masonry member. "Säule" signifies a support with a circular section referencing its wooden prototype. The I-section column of the Farnsworth House implies a rectangle and is in this sense clearly a "stütze." At first glance this also seems to be the case for the cruciform column of the New National Gallery, because its flanges imply a square. But Mies labels the construction drawing "Säule mit Details." His plan of the member cuts through the circular pinned connection in which a ball that is attached to the column sits in a socket that is attached to the roof. This joint is the one location in which the column can truly be called a "säule." The small diameter of the pin represents the compression capacity of steel. It therefore embodies the traditional function of a column, of resisting a vertical load. A classic "säule" is condensed in the detail of its connection to the roof. The cruciform section of the shaft below derives its shape from the untraditional stiffening function in this modern column that also resists lateral forces.

The eight columns are moved away from the corners of the roof. The resulting cantilever has a positive effect on the stresses in the structure of the roof: by pulling downward, the corners lift the mid span upwards, reducing its bending moment and resulting deflection. In plan, the columns stand in an octagon. When combined with the four corners of the glass-enclosure they imply a full circle. This repeats the shape of the pinned connection, which is structurally and conceptually the most condensed part of this steel architecture, in the overall constellation of the building.

The circle in the building plan is only implied and cannot be seen as a full figure. Similarly, the circle of the pinned connection can't be seen because of its small size relative to the column shaft below and roof structure above. In classical architecture a column capital emphasizes the connection between column and roof with elements that project out beyond the roofline. In contrast, in the New National Gallery the connection is set in, inverting the traditional

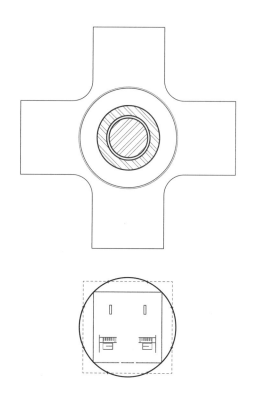

principle. A "negative capital" makes the roof float above the column. Mies reconciles the classic detail that dramatizes the load transfer from roof into column with the modern desire to defy gravity with technology and suspend architectural form in space.

NEGATIVE COLUMN

The negative capital of the New National Gallery is created in the context of a column that occupies the space on the terrace as a bold freestanding object. In the Farnsworth House the whole column becomes a negative volume through the detailing of the glass wall. The plate glass that shelters the interior runs along the edge of the platform. A rectangular steel bar creates a frame to which the glass is secured with a pair of steel flats. Mies uses an invisible weld-joint for the

The column constitutes the only interruption in the continuous view of the river shore. The way in which it meets the glazing is one of the most inspiring details of the house. Two L-sections are welded against the inside flange of the column to hold the glazing frame. They create a pocket of space that goes from floor to ceiling. This makes the column appear as a cavity between the window frames. It implies a conceptual solidification of the exterior. Rather than a solid column standing within the open panorama of the landscape, this "negative column" is an emptiness that cuts apart two volumes of space that come from the exterior and protrude into the domestic space. The detail gives the interior of the house the character of an absence within the landscape. Rather than an object standing in the landscape, the house objectifies the landscape in itself.

inner steel flat that catches the eye of the inhabitant. He places the screwed joint on the less visible outer side. Since welded joints are permanent, a screwed connection is necessary on one side of the framing, in order to replace broken panes. The glass has to be replaced from outside the house. It is not a material that belongs to the house itself, but rather an embodiment of the surrounding landscape that is projected onto the steel frame.

The arrangement of glazing panes cut apart by the cavities of the columns makes the glass wall read like a series of opaque screens that have landscape paintings on them, as often found in traditional Japanese screen walls. They feature stylized images of nature that are continuously painted across separate colorful screens, which are connected with hinges. The gaps between the panels seem to be cut out of the panorama. In the same manner, the negative column on the inner elevation of the Farnsworth House exposes a white surface that seems to exist beyond the landscape. This flattens the landscape to a painting. The entire world is condensed into the surface of the house, and outside of the house remains only a white emptiness.

3.34 The column locations and corners of the glass enclosure imply a circle in the plan of the New National Gallery. This repeats the shape of the pinned column connection in the overall constellation of the building.
3.35 A ball that is attached to the column sits in a socket that is attached to the roof. This pinned connection creates a "negative capital" that makes the roof visually float above the column.

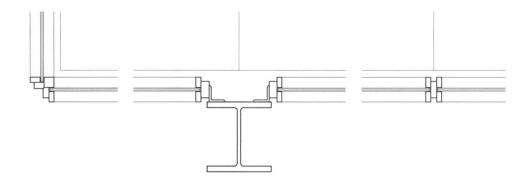

WINDOWS WITHOUT WALLS

Rather than becoming a flattened landscape, the space of the city flows uninterruptedly through the transparent exhibition hall of the New National Gallery. As in the Farnsworth House, the glazing is held in place by steel flats that transfer the wind loads up into the roof and down into the base. But the glass does not coincide with the edge of the steel roof. Mies had been disappointed about the solid appearance of the Seagrams building in New York caused by the reflections in the glass curtain wall. By setting the glass back from the edge of the roof, he prevents such reflections in the New National Gallery, creating a seamless spatial continuity between outside and inside.

If the windows of the Farnsworth House condense the surrounding landscape into a painting, the paintings that are exhibited in the New National Gallery expand into windows that look beyond the surrounding cityscape. The artwork is displayed in a way that is integral to the structural system of the exhibition hall. Paintings are shown on panels that are suspended with steel rods from the underside of the roof girders. Since the roof is entirely supported on columns, the walls can hang down from it, inverting the traditional hierarchy of structural logic and spatial organization. The role of the wall is redefined to exclusively suspend art in space. An exhibition of Mark Rothko paintings in the 1970's exemplifies the spatial power of this concept. In this show the hanging wall pieces are not only the structural armature for the paintings, but also their picture frame. The paintings look like windows cut into these panels, and the exhibition becomes an experiment on how to make windows in an architecture that does not have walls. Traditionally the wall is the architectural element that characterizes the identity of an interior. It also provides the necessary surface to create an aperture that defines the relation of a space to its outside. Since the entire exhibition hall is designed as an outside already, the paintings become windows to an imaginative landscape of self-reflection.

STRAIGHT AND CURVED

The linear expression of the Farnsworth House is an immediate rendition of the straight steel profiles from which its

3.36 (opposite) For the glazing of the Farnsworth House, rectangular steel bars create a frame that stands inside the structural frame of the columns and girders.
3.37, 3.38 A "negative column" between the glazing frames cuts apart two volumes of space, like the gaps between the panels of a Japanese screen wall. This flattens the landscape to a painting.

framework is assembled. The straightness that is expressed in the members of the New National Gallery on the other hand is the result of complex engineering methods. The graceful lines of its columns and roof are produced with steel segments that are curved and tapered. This is apparent in the tapering column shaft that grows slimmer as it rises. While corresponding to the higher stresses in the bottom of the member, the tapering also makes the column appear more vertical than if it was a purely straight extrusion.

It is not immediately apparent that the roof is also assembled from trapezoidal segments. The thin structural grid would deform under its own load if it was constructed as a straight member, since it is only supported in a few points. The cantilevering corners would hang down and the center would sag, creating a depression over the center of the hall. In addition, the heat of the welding causes a further shrinking of the jointed steel elements called *weld-creep*. Mies uses a built-in curvature called *camber* to compensate for all these deflections. This upward curvature follows the exact reverse line of the anticipated deformation. During the construction of the steel grid, a parabolic dome with a height of three feet rises in the center of the square and each corner is lifted up one and a half feet. The roof bends down under its own load and assumes the desired alignment once it is supported on the eight columns. To maintain the desired thinness of the structure, the deformation is accepted, calculated and factored into its design. Without this camber the roof

Steel Framing / Case Studies

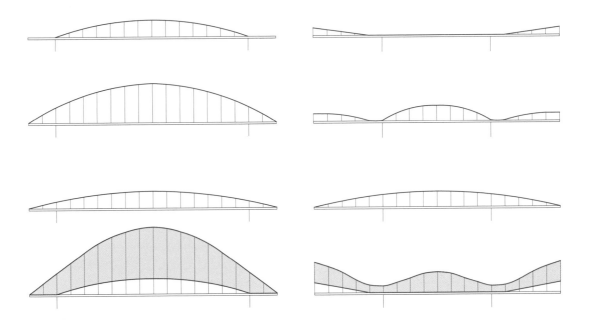

would have needed a much deeper section to come up with the necessary rigidity for the large span to be straight. Yet, even in the final state the roof is not mathematically straight: A visual camber compensates for the sagging optical effects of straight lines with a slight upward bend that is added to the load camber and weld creep camber.

3.39 The walls for the display of paintings hang down from the roof, inverting the traditional hierarchy of structural logic and spatial organization. The paintings look like windows cut into these panels, and the exhibition becomes an experiment on how to make windows in an architecture that does not have walls.
3.40 During the construction of the thin steel roof, Mies uses a built-in camber curvature of the center girder (left) and of the edge girders (right). This curvature compensates for (from top to bottom): the anticipated deflection of the roof under load; for the weld creep; for the sagging optical effect of straight lines on the human eye. Once supported on the columns, the roof bends down into the desired alignment that is visually perfectly straight.

SPACE TECHNOLOGY

The camber can only be designed through a precise calculation of the roof's deformation. The rigid connections of the structure that transfer all forces in all directions increase the variables in the calculations of stresses, material thicknesses, and camber dimensions exponentially. As a result the roof is *statically indeterminate* to a high degree. Mathematical *force-groups*, common in the design of aircraft hulls, were introduced into the structural analysis. Mies and his structural engineer Gerhard Richter used electronic computers that were originally developed for space technology to calculate this complex system. Prior to the introduction of such computers, indeterminate structures had to be engineered with *graphical approximation methods*. These kinds of calculations would not have been sufficient to solve a structure as complex as the roof of the New National Gallery. From the early 1960s

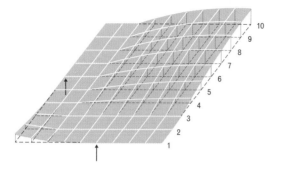
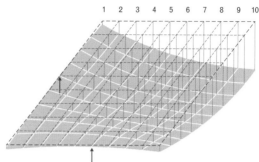

on, engineers wrote specific programs for their calculations in a simple FORTRAN computer language. In the late 1960s when the New National Gallery was built, researchers at the Massachusetts Institute of Technology had developed the general purpose structural analysis program STRUDL, the model for all subsequent software in this field. Any structure that could be modeled as an assembly of two-dimensional elements connected at nodes could now be analyzed with precision.

The construction of the New National Gallery exemplifies the first time in history that such digital technology significantly influenced the creation of architectural form. Computers are used to calculate a complex camber curvature, which is geared towards realizing something that is perfectly level. This constitutes a new relation between the design of a building and its physical presence. What is digitally rendered as a proverbial computer "blob" becomes a stereometric "straight blob" as a constructed result. Upon completion of the roof, the actual deformation of the steel structure deviated only a maximum of three quarters of an inch from the calculated assumption.

Space technology had been the driving force behind the development of the digital computers that were necessary for this engineering concept. And as much as the digital calculation of the New National Gallery roof challenged established modes of design, its erection on site literally lifted construction into a new dimension.

Since automated shop welding is more precise and efficient than manual welding on site, as much of the roof as possible was preassembled in workshops. A rigorous inspection and approval system, as is common in airplane construction, ensured the accurate fit of the various parts that were produced in different locations. The welds were inspected with x-rays to guarantee a consistent structural behavior of all joints throughout the system.

For the purpose of its prefabrication, the square shape of the roof was subdivided into segments suitable for transportation and installation. In order to minimize tensions in the welds that had to be made on site, the final jointing lines between the prefabricated segments had to be straight and run continuously through the entire structure. This informed the decision to preassemble the girders as linear boxes running in east-west direction. Only the center and edge girders were shop-assembled as individual members because of the odd number of structural axes in the square. These preassembled strips could not be transported as 200-foot-long pieces; they had to be subdivided into three installation segments that were later butt-jointed on site.

Once the pieces arrived on site they had to be moved to their correct position on top of the base story. The load of the enormous steel segments could only be placed directly onto the beams of the concrete slab. Railroad tracks were placed over these beams and the steel segments were carted into their plan position on special transportation lorries. The strips were then connected to each other sequentially from the centerline outwards by inserting top plates and crossing girder

3.41 Electronic computers were used to calculate the visual camber (left) and deflection (right) of the roof grid. These complex curvatures are geared towards realizing something that is in the end perfectly level. What is digitally rendered as a proverbial computer "blob" becomes a stereometric "straight blob" as a constructed result.

3.42 For prefabrication, the square shape of the roof was subdivided into linear strips that were transported to the site in segments and installed from the centerline outwards:

1—The center girder is set in place and preassembled box girders are placed to the left and right of it. Individual steel plates for crossing webs, bottom flanges and top plates are inserted between center girder and the adjacent box girders.

2—The third and fourth box girders are laid out on either side of this completed central part and connected to it.

3—This sequence is repeated with four additional box girders to create the surface of the entire roof.

4—Finally, the edge girders are mounted along the square perimeter of the structure.

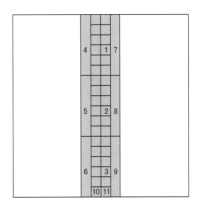

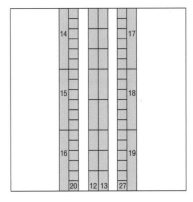

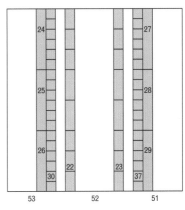

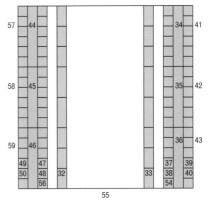

segments between them. Once all pieces of the roof had been welded together, Mies removed the traces of the directional strip assembly: he grinded down all visible butt joints in the bottom flanges of the girders to give them an equal appearance in both directions.

It would have been difficult to assemble the roof in its final height; this would have necessitated a vast shoring platform. It also would have been risky to release the load of the roof step by step from temporary supports onto the final columns. Irregular load distributions during this process could have overstressed the steel structure of the roof and the concrete structure of the base. For these reasons the roof was assembled on the base and lifted up into its final height as one piece.

Hydraulic jacks were placed on temporary lifting towers over each column location to pull the enormous steel square up. The columns were attached flexibly to the edge girder to swing into position as the roof was lifted over the course of a single day.

This installation sequence is different from previous building procedures. Construction becomes a precisely calculated operation that replaces a gradual process. A traditional masonry building grows slowly over time, brick-course by brick-course. The lifting of the New National Gallery roof has more in common with a rocket launch. In space technology, extensive computer calculations guide the development of new machines and materials that make the extraterrestrial journey possible.

The technological preparations that take many years culminate in a countdown and a single moment of lift-off that only lasts for a few seconds. After the energy-consuming launch has been successful, a state of suspension beyond gravity is reached and the rocket is on an orbit that follows cosmic rules of centrifugal force and speed.

3.43 The prefabricated linear segments were site-welded together on top of the base. The center girder consists of a web and bottom flange. The picture shows the upward curvature of its built-in camber.
3.44 Every second bay of the roof grid is shop-welded as a box girder. On site these linear steel boxes are connected with individual webs and flanges to create the full roof structure.
3.45 (overleaf) Upon completion of the assembly, the roof was lifted up into its position over the course of a single day.

In the New National Gallery a similarly new type of "space technology" is employed to produce a civic space that is sheltered under a floating steel roof. The rocket Apollo 11, named after the god of the arts, went out to visit distant planetary spheres and allowed mankind to experience the Earth rising above the horizon of the Moon in the same year in which the Gallery for the Art of the Twentieth Century was completed. The same technologies that were used to control the forces that projected mankind into outer space made it possible to overcome gravity architecturally. And the structural elements of this architecture are placed in a spherical and hence planetary constellation.

Farnsworth House & Gallery of the Twentieth Century

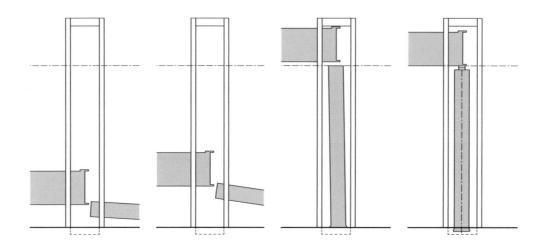

Steel Framing / Case Studies

ORDERS

While the most advanced materials, jointing techniques and engineering concepts are used to design the Farnsworth House and the New National Gallery, their steel members allude to spatial archetypes from the origin of architectural creation. In his search for the principles of modern building, the classic orders of Greek antiquity provide Mies with a system of references that ground his floating steel elements in time. The visually voided connection between

3.46 Lifting sequence (from left to right): The column is connected to the underside of the roof. First the roof is lifted from its temporary props on the base. Then it is pulled up to a height slightly above its final elevation and the columns are secured in place. Finally the roof is set down onto them.

3.47 Hydraulic jacks on temporary lifting towers over each column location pull the roof up. During this lift the column swings into place. This construction sequence is similar to a rocket launch: extensive computer calculations and mechanical preparations culminate in a countdown and a single moment lift-off.

3.48 The features of the Greek orders were derived from translating timber architecture into stone. They tell a story about the formal origin of the building components in another material. The "female" Ionic order (right) has slender columns and an uninterrupted frieze as can be seen in the Erechteion on the Acropolis of Athens. The "male" Doric order (left) looks more grounded and heavy, as exemplified in the Parthenon.

column and roof of the New National Gallery can only be recognized as a negative capital, because the historic concept of the capital as a dramatized architectural moment is embraced as such. In the orders of the Greek temples, such features were derived from translating timber architecture into stone. The detailing tells the story about their formal origin in another material, rather than only representing their current physical dimension. They incorporate the wooden narrative of this story as much as the stone language in which it is told.

In the process of this sculptural "petrification," the orders make an already existing canon of architectural elements harder and more durable, transforming physical material into artistic matter. The Farnsworth House and the New National Gallery are in a direct lineage with this artistic development, further advancing the process by translating stone into steel. Their steel columns and girders embody the linear qualities of the original timber as well as the sculptural technique of its later interpretation that was carved in marble. Yet they are assembled with welded connections that deny either of those references.

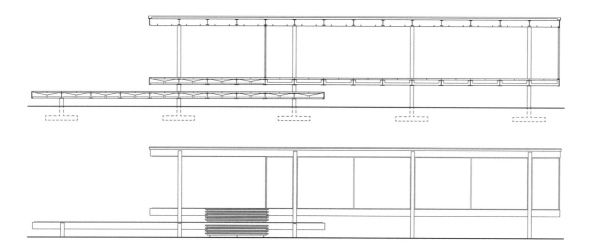

Their framework resolves the ancient paradox of a wooden architecture that is built in stone through the distillation of both materials in a third: steel. This allows them to incorporate all traditional materials – and create an architecture beyond any particular materiality.

Like an ancient Greek temple, the rectangular proportions of the Farnsworth House celebrate the wooden tectonics of a building that is constructed from columns and beams. Two rows of columns run in parallel. In each row the individual columns are connected with a principal girder, the *architrave* of a classical order. Secondary beams span between the two principal girders. This configuration is entirely one-directional. A conceptual problem that the ancient stonemasons encountered in the *frieze* of the Greek temple persists in its twentieth century reincarnation: at the short sides of the house the edge of the roof is framed with the same C-section as on the long sides. But it does not span between the columns, and therefore it does not have the same structural function. In a pure construction logic, the short side of the elevation would show the side of a secondary beam spanning between the two principal girders, rather than the girder itself turning the corner.

The ancient solution to this artistic problem is to wrap the architrave around the corner and use the side elevation for the front elevation too, although it is structurally meaningless. In the Farnsworth House, Mies solves this issue the same way. He respects the ancient tradition by continuously wrapping the edge girder around the corner. At the same time he also celebrates modern principals of honesty and abstraction: in the longitudinal section drawing the C-section of the girder reads as half an I-section that is generated by cutting through one of the potentially infinite number of secondary beams because the house ends. Although the member is not actually subtracted from a larger profile like the custom L-section of the fascia piece, the context of the assembly suggests that it is conceptually only half a beam, with the other half being cut off.

IONIC AND DORIC

While both buildings represent the idea of the orders in general, specific features of the Farnsworth House and the New National Gallery reference particular kinds of orders that relate to their program and meaning. The slender columns and narrow uninterrupted frieze of the Farnsworth House read as a twentieth century homage to the *Ionic order*. In Greek antiquity this "female" order was predominantly used for the sanctuaries of goddesses. Thus it seems appropriate that it finds its modern incarnation in a house built for a lady, Edith Farnsworth. The "male" *Doric order* that looks more grounded and heavy is referenced in many features of the New National Gallery. The tapered shape of its column is a nod to the *entasis* of the Doric column. This aesthetic device was used in antiquity to support the visual gesture of tallness and express the resistance of the column against vertical load. The entasis also related the marble column to its timber prototype. In the earliest temples the column had been a tree trunk with a naturally diminishing diameter. Because of the hardness of its wooden ancestor, the Doric marble column stands without a base detail directly on the stone foundations. The stone pavers of Mies' terrace create a similar visual effect; they conceal the steel base plate connection to the concrete substructure. Along the perimeter of the roof the prominent vertical stiffening plates read as an aesthetic homage to the *triglyphs* in the Doric frieze that resemble beam-ends. They subdivide the face of the edge girder

3.49 Mies' steel buildings are in a direct lineage with the Greek orders. Their steel members embody the linear qualities of the archaic timber temple as well as the sculptural technique of its later classic interpretation that was carved in marble. The slender columns and narrow uninterrupted frieze of the Farnsworth House read as an homage to the Ionic order. In the longitudinal section, the C-section of the edge girder reads as half an I-section that is generated by cutting through one of the potentially infinite number of secondary beams.

3.50 The vertical triglyphs in the Doric order resemble beam ends; they subdivide the frieze into rectangular metopes, plates that hold the sculptures of gods and heroes.

into a series of distinct steel plates that accordingly reference the *metopes* of this kind of frieze, which hold the sculptures of gods and heroes. Since the roof of the New National Gallery has a structural depth of six feet, its blank steel metopes refer to the human body as a potential sculpture that could inhabit this frieze of the temple of modernity.

WHITE AND BLACK

The complementary dialogue between the Farnsworth House and the New National Gallery can also be seen in the different colors that Mies applies to their steel members. In order to protect structural steel from corrosion, its surface needs to be treated with a protective coat of paint. Mies takes this necessity as an opportunity to strengthen the abstract quality of the artificial material that can only display its "natural" rust color during the process of its speedy deterioration. The most extreme hues white and black pay artistic duty to the totality of each spatial concept. White stands for absence. It emphasizes the idea that the architecture of the Farnsworth House is the result of a reductive process. The elements of the house are physically cut away from the steel sections just as the entire house is conceptually cut out of the landscape. Rather than generating space its members take away from space. In a logical constructive manner the house subtracts the world from itself. On the contrary, black is the condensation of all colors. It emphasizes the idea that the architecture of the New National Gallery is assembled from many individual steel plates that converge to form a bold object, which reconfigures the city and collapses the history of architecture onto itself.

Like no other buildings the Farnsworth House and the New National Gallery are the crystallization of the aspirations of the modern movement. They embrace the most advanced technological possibilities of their time both in design and execution. Completed in a time when mankind started to fly to the moon, they represent the enthusiasm of an era that believed in limitless scientific advancement and technological progress. At the same time they embody the most primal gestures of spatial creation and pay homage to the earliest moments of architecture, integrating modern achievements with traditional concepts of beauty and grace. Their features provoke multiple readings that escape an intellectual understanding. In an analysis of their steel elements, each discovery is immediately complemented by another interpretation, broadening references across scale and time. These references are distilled into the linear elements of the steel structure that abandons any stylistic features in order to emphasize the universality of architectural space. It is as if the densest construction matter inspired Mies to create architecture with the densest meaning.

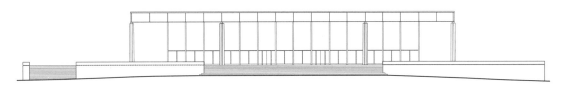

3.51

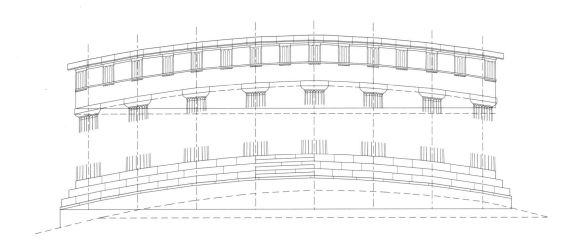

3.52

3.51   The Doric order is referenced in many features
of the New National Gallery: The tapered shape of its
column is a nod to the entasis of the Doric column;
the vertical stiffening plates along the perimeter of
the roof read as an aesthetic homage to the triglyphs
in the Doric frieze.

3.52   Base and roof of the Parthenon on the Acropolis of Athens are built with an upward curvature that
compensates for the sagging optical effects of long
straight lines, just like the visual camber in Mies' roof.

3.53 (overleaf)   A new type of "space technology"
produces a civic space that is sheltered under a floating steel roof.

SEAMLESS STEEL TUBES

3.54

The profiles of I-beams originate from the production of railroad tracks; hollow section types were first manufactured as pipes for plumbing and mining. They have an inherent stiffness due to their geometry and were later reappropriated for the use as load bearing members in steel structures. Such *hollow structural sections (HSS)*, also simply called *tubes*, have a consistent wall thickness; they do not consist of major and minor parts such as the flanges and webs of rolled I, C and L sections, which are designed to resist a specific kind of load. This makes tubes versatile building components, capable of resisting different types of stresses. In particular, circular tubes are literally all-around in their use. They are equally strong in all axes of their circumference, and loads can be applied to them in manifold angles.

Traditionally steel tubes were produced by bending sheet metal into the desired hollow shape. This resulted in a continuous seam that created a weak line within the homogenous surface of the material. The invention of a production process called *rotary piercing* by the German industrialist Reinhard Mannesmann in the beginning of the twentieth century eliminated this inconsistency. This process made it possible to create thick-walled circular tubes that are completely seamless. A heated steel cylinder is fed between two tapered rollers, which are rotating in the same direction and

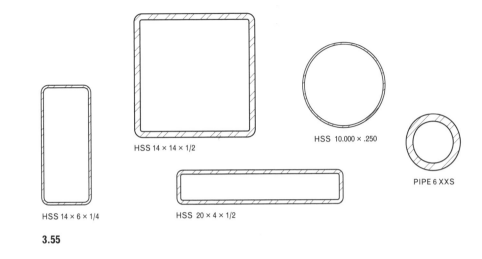

HSS 14 × 14 × 1/2

HSS 10.000 × .250

HSS 14 × 6 × 1/4

HSS 20 × 4 × 1/2

PIPE 6 XXS

3.55

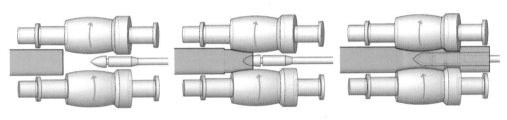

3.56

compress the material. The resulting friction creates specific stresses in the very center of the steel cylinder; they cause a longitudinal void to form at its axes. Then a pointed metal bar called *mandrel* gets inserted into this stress-created central void. It forces the material further outward from within the cavity and presses it against the tapered rollers;

this sets the inner and outer diameter of the resulting tube. The tube is then cooled and can be cold worked for refinement. Additional thermal treatment can increase its toughness and strength. Finally, the tube is straightened through rotary equipment and cut to the desired length.

Their stiffness makes tubes ideal for use as columns or more generally members that require a strong resistance against buckling. However, since their surfaces are curved and cannot be reached from two sides, they are difficult to connect into a three-dimensional framework; hence they require complex jointing techniques.

3.55 Section profiles of square tube, structural rectangular tube, circular tube and thick-walled circular tube
3.55, 3.56 In rotary piercing, a raw steel cylinder is pushed between two rotating tapered rollers; the stresses induced by the rolls cause the center of the steel work piece to fracture and become hollow; then a pointed steel bar gets inserted into this central cavity to shape a uniform inner diameter for the resulting steel tube.

Seamless Steel Tubes

SENDAI MEDIATEQUE

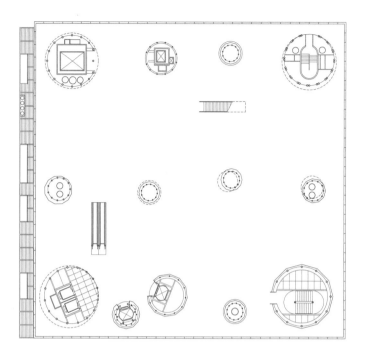

The design of the Sendai Mediateque in Niyagi, Japan explores the structural character and spatial presence of the circular steel tube just as the Farnsworth House and the New National Gallery are inspired by the shapes of rolled and composite steel sections. The building was completed in 2000 by the architect Toyo Ito and the engineer Mutsuro Sasaki. It is a center for culture and media that redefines the program of a library for the communication technologies of the digital age. Thirteen circular shafts, which are constructed as a transparent network of steel tubes, support seven floor plates that are placed at irregular heights. Each floor houses a particular type of media collection. A glass facade covers all four sides of the building. This facade is open

at the corners, emphasizing the openness and theoretically unlimited expansion of the spatial concept.

The shafts of the Sendai Mediateque render the spatial character of the steel tube in architectural scale: their perimeter carries the building weight like a structural tube and their cavity supplies the building program like a plumbing pipe. Their linear void facilitates vertical movement and distribution: some contain circulation such as stairs and elevators. Others enclose mechanical systems. A few are left empty, awaiting their future content yet to be invented. As the shafts warp their way up in the building, they change their diameter and configuration. This creates places with distinct character in the open and fluid spaces

　　　Steel Framing / Case Study

of the floors. The shafts establish nodes of activity without compartmentalizing closed areas.

SHAFT STRUCTURE

Though they may appear subtle and transparent, the shafts are extremely rigid and strong. Their different sizes and positions reflect specific tasks in the overall structural system of the building: four large shafts with diameters of up to 30 feet resist horizontal forces from wind and earthquakes. These main shafts are located in the corners of the plan to avoid torsion in the floor plates. If they were placed in irregular locations, the floors would have the tendency to rotate around them under lateral pressure and destabilize the building. They are built as a triangulated lattice tube structure that is completely stiff and cantilevers up vertically from the building basement. In the first lower level they connect to another kind of steel frame that is completely flexible. It stands with a needle like support in cast iron pans on the concrete substructure below. In the event of the ground shaking back and forth during an earthquake, the rigid structure of the upper shafts slides as a whole on this ductile frame at the foundation,

3.57, 3.58 The structure of the Sendai Mediateque is a transparent network of steel tubes. Thirteen circular shafts carry the floor plates and contain the building infrastructure.

3.59 (overleaf) The shafts have different sizes and configurations, creating distinct places in the open and fluid space of the floors.

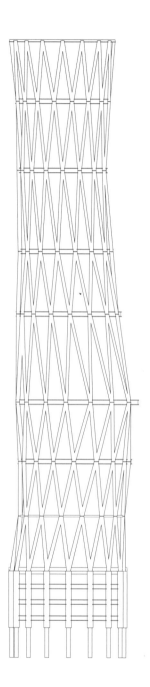

which acts like a buffer to the ground to absorb all the stresses. This arrangement limits any structural damage to the lower level of the building.

The nine smaller shafts that are only up to seven feet wide do not resist horizontal forces. They are arranged in plan solely as supports for vertical loads and their structure is not triangulated. Instead they are constructed from parallel vertical tubes with intermediate ring shaped *hoops* that prevent the individual members and the shaft as a whole from buckling. The overall stability of these smaller shafts against buckling is further increased by the twisting of their shape. This disperses the reaction from floor to floor and makes the shaft behave like a metal spring that is stronger when under tension.

All the shafts are composed of thick-walled tubes, ranging from 140mm to 240mm in diameter. Just as the individual tubes surround a circular void, so too do the larger shafts in architectural scale. In the construction drawings for the assembly of the shafts, the tubes are not shown as a physical thickness. They are instead only indicated with their center-lines as vectors in a dynamic network between multiple points of connection. Their diameter registers only in the places where they attach to the floor plates, just as the overall perimeter of the shaft is only shown where it cuts through the floors. This emphasizes the immaterial idea of the nodes rather than the connecting members themselves. I-beams, C-profiles and L-profiles are composed of planes. The proportions of their section, as well as the thickness of flange and web express the magnitude and type

of force they resist. Tubes do not reveal the wall-thickness of their metal and therefore give no evidence of their physical capacity. They appear as an abstract line of force in space. This makes them ideal components for the creation of an architecture that does not want to represent any solidity or mass whatsoever.

HONEYCOMB SLAB

When moving up and down through this dematerialized steel tube latticework, the visitor experiences how the shafts pass through circular holes in the floors without interrupting their dynamic expression. Each of the square floors measuring 164 foot by 164 foot becomes an abstract flat plane cutting through space. The slabs are constructed to be as thin as possible, spanning up to 66 feet between supports with an overall depth of only 18 inches. Their weight is also minimized in order to allow for the structural thinness of the vertical shafts that carry them.

To minimize thickness and weight, each floor consists of continuous top and bottom steel plates, which are connected to a *honeycomb slab* by a system of vertical webs. The arrangement of these webs corresponds to the different stresses that occur across the floor. Their geometry creates a transition between the circular perimeter of the shafts and the rectangular floor grid: around the shafts, the webs connect in a radial arrangement to the individual tubes of the latticework. A spider-web-like pattern diverts these lines to meet a rectangular grid with a regular three foot spacing that fills the remainder of the floor area.

3.60 The large shafts are built as a triangulated tube latticework that resists horizontal forces. This rigid structure stands on a flexible steel frame in the basement, which absorbs forces from earthquakes.
3.61 The small shafts are constructed from parallel tubes with intermediate ring shaped hoops.

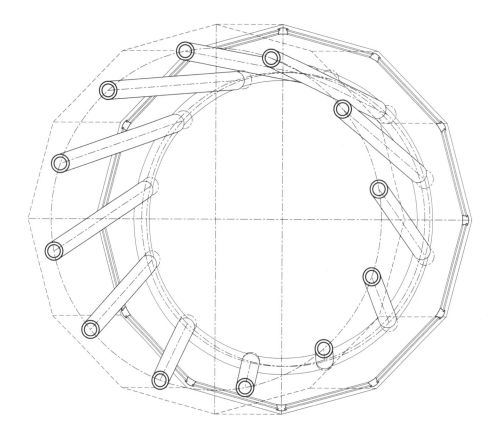

The result of this complex arrangement of pieces is an economical steel floor slab that is dynamically rational and has been reduced to its bare essentials. Even the thickness of the steel used for its assembly varies to accommodate different stress levels. Around the shafts where the stresses are particularly high it is 16 to 25 millimeters thick; further away from the shafts where the stresses are lower it is only six to twelve millimeters thick. The roof of the New National Gallery is composed of a similar arrangement of steel surfaces that vary in grade and thickness to address different stresses in a uniform grid. The open arrangement of intersecting bottom flanges reveals the organization of the webs and the depth of the slab. In contrast, at the Sendai

Mediateque the structural characteristics of the floor do not become architectural features. The spectacular internal organization of the floor plate is hidden from our view in the completed state. Although the spider-web like configurations of the steel pieces around the shafts are suggestive of the spatial draw that these nodes of activity create on the floors, Ito does not utilize this to create a formal expression. The elements of the honeycomb floor exist solely to fulfill the purpose of minimizing thickness and weight.

SEAMLESS

The welded connections in the slabs and shafts of the Sendai Mediateque aim to eliminate the visual presence of joints to emphasize a dematerialized and abstract

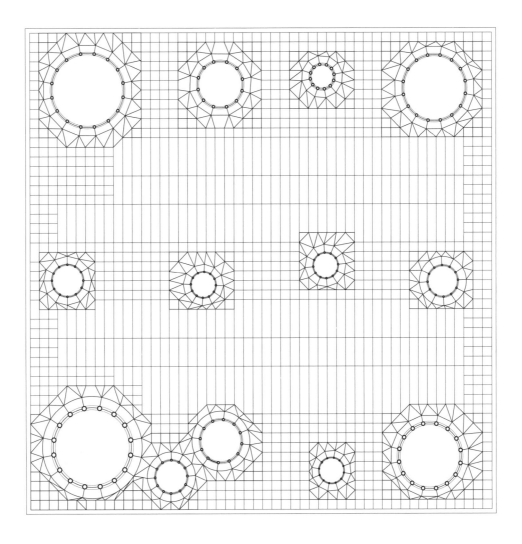

3.62 The shafts are composed of thick-walled tubes. The twisted shape in which the tubes are assembled increases the stability of the shaft.

3.63 Each floor consists of continuous top and bottom steel plates which are connected to a honeycomb slab by a system of vertical webs.

3.64

3.65

3.66

appearance. The linear welds between bottom pieces of a composite steel slab would typically require *backing material* underneath: a steel bar that covers the groove joint from below, preventing the molten metal from dripping down through the gap between the plates. Such a backing interrupts the visual continuity of a steel ceiling when seen from the below. In the Sendai Mediateque, to avoid this an *overhead-weld* was made from underneath the joint line to create a seamless surface. This kind of welding is extremely difficult, since the joint is worked on against gravity. The welder needs to carefully adjust the voltage and resulting temperature, as well as the amount of filler and the speed of motion, in order to make sure that no excess molten metal drips down. While rarely used in architectural construction, this type of welding is quite common in ship building on the inclined surface of a steel hull. Welders from the shipbuilding industry were hired to assemble the honeycomb slab of the Sendai Mediateque.

In assembling the shafts, virtually seamless weld joints between the tubes resonate with the seamless fabrication method with which the tubular members themselves have been manufactured. In order to transfer loads properly to their joints, the central axis of each individual tube has to intersect in a single point so as not to create any eccentricity. For this purpose the end of each tube needs to be cut in complex curves that are determined by how it meets other tubes in the joint. Along these curved lines, the tubes were fitted together with tight gaps and jointed

Steel Framing / Case Study

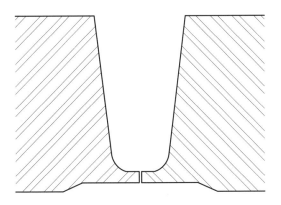
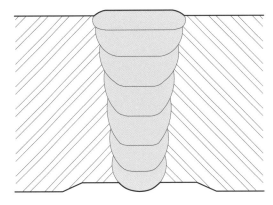

with *narrow gap welding*: a computer-operated machine mills a precise profile with small angles in the range of two to twenty degrees onto the faces of the two members that will connect. These inner faces of the joint become almost parallel, leaving only one side slightly open to place the weld. Because of the limited accessibility to the bottom of the gap a special narrow welding torch is needed. This type of welding requires less metal filler and increases precision in the joint since distortions that can be caused by heat are reduced. The appearance of the weld seam is minimized, creating a visual continuity of the tubes across the connecting shafts of the Sendai Mediateque.

3.64 The steel pieces that compose the bottom plate of the honeycomb slab are jointed with overhead welding to create a seamless ceiling surface.
3.65 The arrangement of the webs inside the slab creates a transition between the radial logic of the shafts and the rectangular floor grid.
3.66 The tubes of the shafts are assembled with narrow groove welds that minimize the size and appearance of the weld seam.
3.67 For narrow groove welding, the sides of the joint are almost parallel with a refined profile.

The steel latticework becomes a gigantic woven net, rather than a tectonic assembly of individual pieces.

The connection between shafts and floors resembles a fitting between different sections of a plumbing pipe: the upper steel surface of the floor plate rests directly on a ring that connects the individual tubes of each shaft. Bolted connections transfer horizontal forces from the plate into this ring, but no moment is induced. This structural detail avoids eccentric bending stresses, which might be harmful to the steel tubes of the shafts. Both the latticework of the shafts and the honeycomb slabs of the floors were shop welded in a factory up to maximum transportable size and then installed on site, floor by floor. By not fixing ring and plate during the assembly, the problem of constraining stresses incurred through welding heat was eliminated. Upon completion of the steel work three inches of lightweight concrete was cast on top of the steel floor plates. The concrete is structurally integrated by means of *shear connectors* that prevent sliding on the slick steel surface.

3.68

3.69

A layer of heat resistant glazing envelops the shafts on all floors. Trapezoidal glass segments accommodate the irregular shapes required in each specific location. The glazing does not fill the spaces within the steel latticework. Rather, it spans from floor to ceiling and only connects once, at mid-height, to the steel tubes of the shaft for lateral stability. Here the intermediate window sash that supports the glass is welded against two steel flats. These are held in position by a steel pipe, which is welded to the metal tube of the shaft. The position of the glazing relative to the structure makes the shafts appear as individual buildings themselves, turning the space of the floors into an exterior experience.

WIRE FRAME

All floors of the Sendai Mediateque have a similar plan. Yet each of them is unique because of the changing diameters and locations of the shafts. The entry level is treated just as the other floor plates. It does not read as a ground, but as just another suspended surface that happens to coincide with the surrounding street level. Rather than being grounded in a particular place, the Mediateque is situated in a specific time in which things are not determined by gravity: the digital age. For this reason Ito inverts traditional tectonics in which vertical members are introduced to serve horizontal surfaces. The floor plates look as if they are created to stabilize the shafts and hold them in

3.70

3.71

their relative position, rather than the shafts looking as if they were erected to carry the floors.

The *wire frame* model of the Sendai Mediateque articulates this hierarchy by emphasizing the shafts. In this kind of computer model, the building shape is defined as a distorted three-dimensional network of flexible "wires" that can be can be bent into any shape. While often used to simply describe extravagant

3.68, 3.69, 3.70, 3.71 The steel elements of the shafts and floors were shop-welded to segments of a transportable size which were then field-welded together on site.
3.72 (overleaf) The built expression of the Sendai Mediateque wireframe resonates with the digital technology it houses.

shapes, the wire frame of the Sendai Mediateque is not only the geometric description of a free form, but a rational representation of its structural concept as defined by *finite elements*.

This structural design method was invented the engineers John Argyris and Ray Clough in the 1950s. In a finite element model, physical members are represented by a number of conceptual "elements" that describe structural behavior. Linear members such as columns, beams or cables are represented as one-dimensional elements: Vectors positioned at their centroidal axis. Physical properties such as axial, bending, and torsional stiffness are mathematically assigned to each element. Represented as

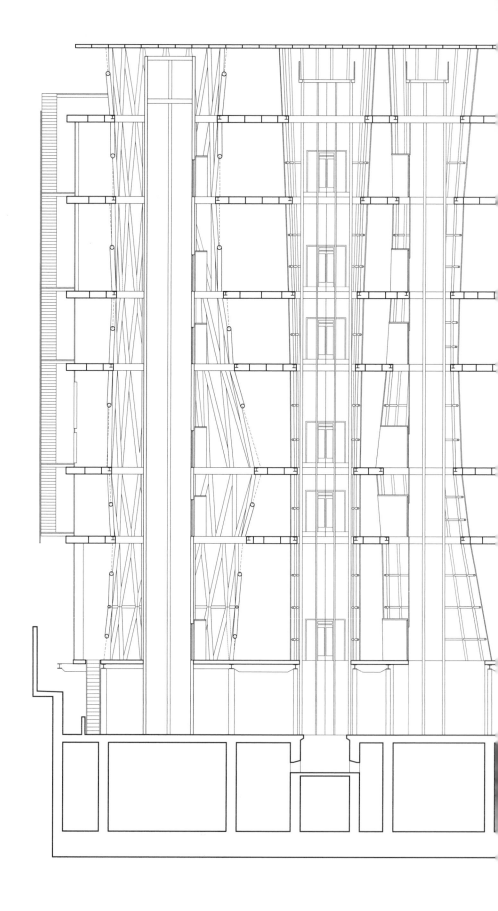

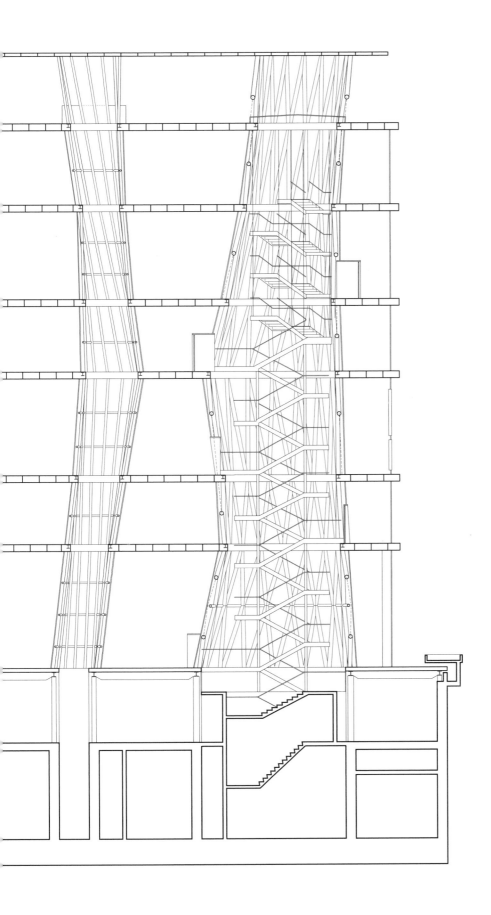

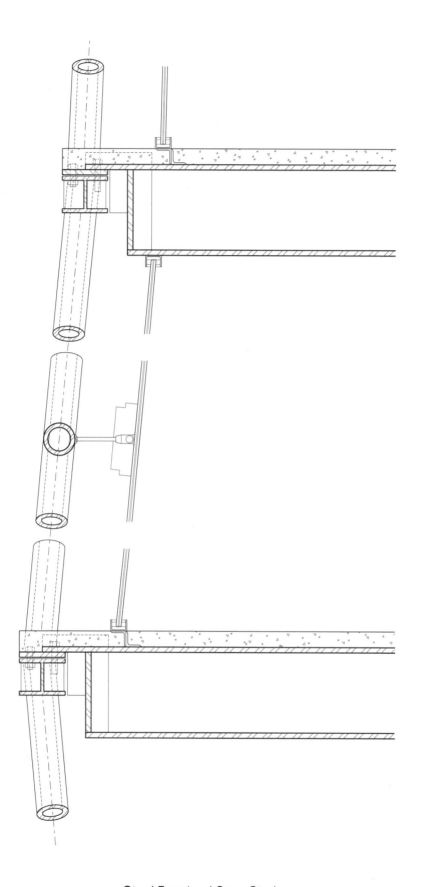

Steel Framing / Case Study

abstracted vectors, they interact with other elements in the model by means of *nodes*: hypothetical points of displacement at the end points of the vector. These one-dimensional elements are well suited to describing trusses, grids and frames.

Surface members like floor plates or shells are represented as two-dimensional elements in the shape of flat or curved triangles or quadrilaterals. They describe stresses for bending or membrane action. In such elements the nodes are placed at the corners; additional nodes can be

placed along the edges or even inside the element for more accuracy. The finite element model is then used to simulate the structural behavior of the building by applying loads to specific nodes. As a result the elements move in a way that is dictated by their defined properties.

The vectors of one-dimensional elements read as an abstraction of linear members connecting to points. The shapes of two-dimensional elements read as an abstraction of a surface connecting to a network of lines. This *reduction of dimension* becomes an architectural reality in the structure of the Sendai Mediateque. The steel latticework of the shafts consists of linear metal members that are assembled to a frame. It becomes a literal construction of the one-dimensional finite elements of the wire frame model. The centerlines

3.73 The top plate of the steel floor rests on a ring of composite T-sections that connects the individual tubes of the shaft. A layer of heat-resistant glazing envelops the shafts, turning the space of the floors into an exterior experience.

3.74 In the wire frame model the floor plates look as if they are created to stabilize the shafts and hold them in their relative position.

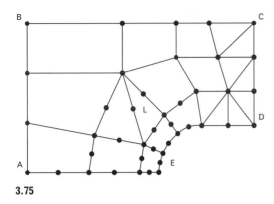

3.75

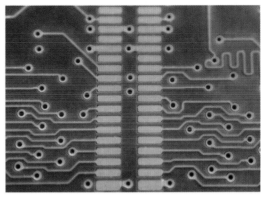

3.76

of the tubes are hollow. In its very core the built structure is dematerialized as the vectors in the model themselves. The steel webs in the floor plate physicalize the finite elements that would describe it as if it was a solid slab. The steel pieces physically manifest the abstract linear representation of a solid building as a transparent structure.

HARDWARE AND SOFTWARE

The built expression of the Sendai Mediateque wire frame resonates with the digital technology it houses: The shafts look like bundles of wires and cables in a computer. The hidden chambers of the honeycomb floor-slab allude to cellular storage for data or energy. Even the welding technology that is employed to connect the steel pieces resembles the soldering that is used to assemble the wires in a micro-processor. In micro-processing chips, hollow nodes are placed on a metal plate in a pattern that resembles the spatial nodes created by shafts on the floor plates of the Sendai Mediateque. Computer *hardware* is

defined as the "platform for the processing of information." The structural components of Ito's building are its hardware; the elements introduced for specific programmatic uses within them become its software.

The circulation systems in the shafts have no structural relation to the steel tube latticework. The elevators run along straight rails that are suspended within the warped tubular space. The stairs cantilever from floor to floor; they "program" the shafts for a specific use. The geometry of the stair behaves like a flexible mass of data that is molded when traveling through the amorphous void. This data consists of treads, rises, landings and width of run, which adapt to varying parameters of the steel cage they inhabit. Different numbers of steps are required to get from floor to floor because of shifting floor heights. The diameter of the shaft also varies, changing the available space for containing these steps; thus the stairs have to be arranged in different numbers of runs. The shaft in the northeast corner of the

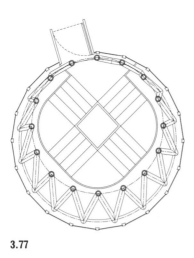

3.77

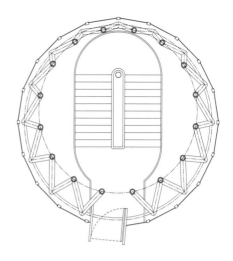

floor plan, for example, shrinks in diameter as it approaches the top of the building. In response, the stair switches from a double run to a quadruple run.

NETWORK

The Sendai Mediateque questions the traditional hierarchy between load bearing and carried members so powerfully represented by the New National Gallery and the Farnsworth House. Mies' buildings challenge archetypal concepts of columns and beams with their large spans, floating cantilevers and seemingly magic connections. The drama of their

appearance is created by how technologically advanced features read against the referential system of tradition. The "negative column" of the Farnsworth House and the "negative capital" of the New National Gallery celebrate the possibilities of steel construction by evacuating the volume from discernible classical features. In the Sendai Mediateque, the corporeality of architectural form is denied altogether. It dissolves the concept that vertical members "bear" horizontal members both structurally and visually. This is replaced by a purely structural "holding up" of the floor plates by the shafts that is not expressed in their connection or figural appearance.

The idea of a steel *frame-work* that captures the world through discrete architectural limits is replaced by a boundless transparent *net-work* that denies stability, scale and permanence. There is no separation between the spaces in which media gets produced and in which media gets consumed. The world becomes a dynamic network of receiving, collecting and distributing data.

3.75 In the finite element method physical members are represented by a number of conceptual elements consisting of vectors and nodes that are sufficient to describe its structural behavior.

3.76 The spatial nodes created by the shafts on the floors of the Sendai Mediateque resemble the pattern in which hollow nodes are placed on the surface of a micro-processor.

3.77 The stair is a software of data consisting of treads, rises, landings and width of run. It adopts to the changing shape of the shafts as it travels up through the building by organizing the treads in different runs.

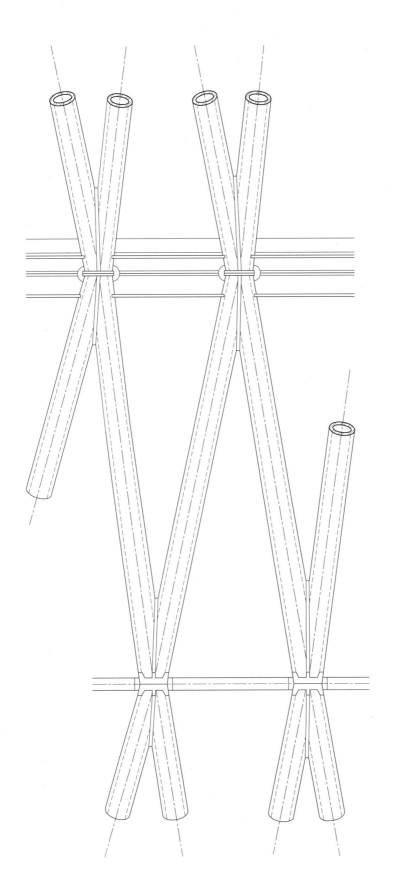

Steel Framing / Case Study

The Sendai Mediateque represents a total environment interconnected through information technology, where the separation between artificial creation and natural phenomena begins to vanish.

All steel members used in architectural construction have an industrial origin and were only later appropriated for use in building design. As such steel has no reference to traditional craft. Rather, it carries associations with other technologies into the creation of form and space. Mies embraces the industrial origin of steel and ennobles it in a transformed classic architecture. He utilizes the visual tension between the raw appearance of the steel sections and the heroic order in which he places them. Ito uses the untraditional abstract forms of steel to dissociate his building from the classic canons. He creates an inhabitable structure that denies tectonic references, and he uses the expert craftsmanship of shipbuilders to execute it. A non-physical expression for a post-industrial age, the Mediateque relates to digital technologies that operate in the microscopic scale of the microchip and the global scale of the world-wide-web.

3.78 The centerlines of the tubes are hollow. In its very core the built structure is as dematerialized as the vectors in the wireframe model.
3.79 The steel framing of the Sendai Mediateque is a boundless transparent network that denies stability, scale and permanence.

COMPOSITE WOOD

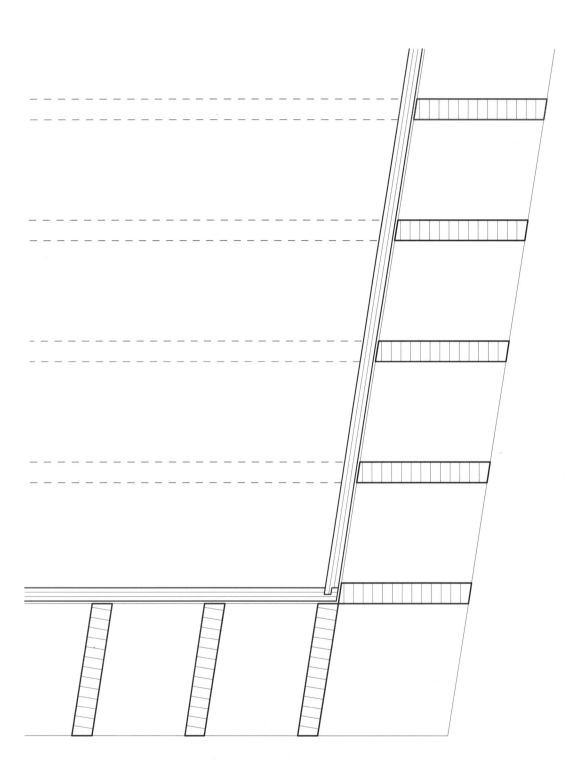

AARAU MARKET HALL

GÖTZ COLLECTION

Composite Wood

COMPOSITE WOOD

The linear shape of a tree and the directionality of its grain have always determined the character of timber buildings. Throughout history, wooden logs were cut into posts and beams. Complex framing techniques were used to assemble them to three-dimensional structures. These frameworks were limited by the dimensions and hardness of the natural wood. Composite wood was invented to meet new demands with old materials. Instead of cutting logs directly into posts, beams or boards, they are sliced into strands, fibers or veneers. This raw wooden mass is bound by adhesives to create a composite member of desired shape and consistency. The structural behavior of this new wood product is enhanced because of its engineering; yet it retains many desirable characteristics of the natural material. It is renewable and easy to work with; it can be jointed in simple ways and does not change the building procedures of site-assembled carpentry. These qualities have made it the top choice for efficient construction in many countries in the twentieth century. The tectonics of woodwork have always been at the core of architectural creation. With composite wood, this tradition confronts mechanization, provoking a distinct dialogue between craftsmanship and modernity.

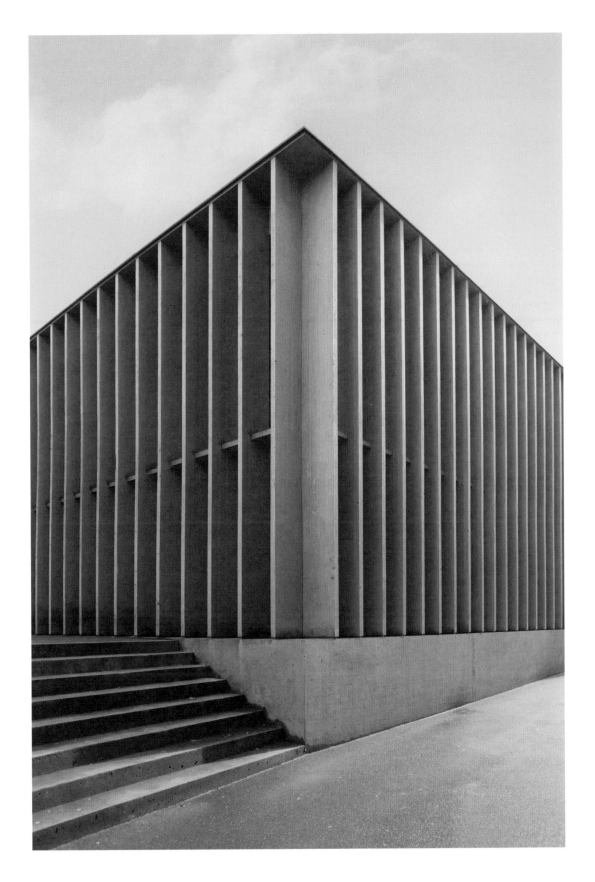

Composite Wood

GLUED LAMINATED TIMBER

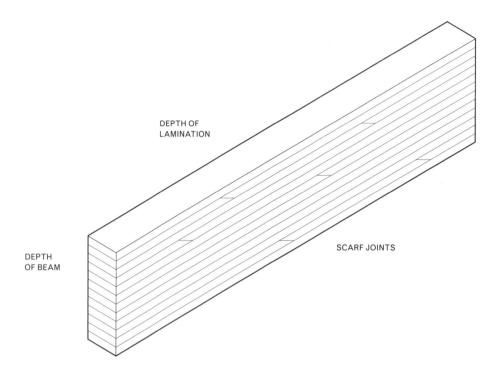

DEPTH OF
LAMINATION

DEPTH
OF BEAM

SCARF JOINTS

The invention of the *glued lamination*, also called *glulam*, revolutionized wood construction by eliminating the structural and dimensional limits of natural lumber. In the beginning of the twentieth century the German carpenter and saw-mill owner Otto Hetzer produced curved eaves made from horizontally arranged wood laminations bonded with adhesive. He used these interlocking timber layers to glue vertical posts and horizontal rafters to a single unit, the *glulam portal frame*. This technology offered an industrial alternative to traditional carpentry work: lumber pieces are glued to a structural frame rather than connected with complex joints. The method was quickly extended to produce the entire member

as a prefabricated glued lamination of wooden layers, in order to control its dimensions and properties. Glulam production was further advanced in Switzerland and the United States. It established an industrialized way of building with timber that is particularly popular in rural regions that have an abundance of wood.

In modern glulam production *scarf-jointed* strips of timber are connected end to end, to form a linear lamination layer. Several such layers are fed through a glue extruder and stacked mechanically, staggering the placement of the joints. This stack is bonded under heat and pressure to form a single member. The strength and quality of this artificial

Composite Wood / Technology

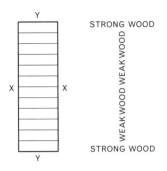

STRONG WOOD

WEAK WOOD WEAK WOOD

STRONG WOOD

timber element is determined by its design. For example, in the laminations of a beam, strong wood can be placed at the top and bottom of a section, like the flanges in a steel beam. Instead of a web connecting the two flanges, weaker wood is placed between them as a fill material. The result is a strong composite member that is more technologically advanced than its appearance suggests.

The individual wood pieces that compose a glulam member can be cut from young trees. This is an economical improvement in heavy timber construction, as big trees are costly to harvest, season, and transport. Now that the size of the timber is independent of the size of the log, a structural beam or column can be custom made to the distance it must span, the load it must carry, and the shape it must take. The factory

equipment is always used in the same way, without differentiating between a standard and a custom member. Even curved or arched shapes are simple to make: thin laminations are bent individually with the required radius and then glued together.

Glulam timber is strong enough to substitute for steel or reinforced concrete in large-scale buildings. Yet it does not require the enormous embodied energy or the upkeep against corrosion of these materials. It is also easier to transport and handle because it is much lighter. During installation it can be tooled, trimmed, and drilled easily. Because of these features, glulam is an ideal technology to create a strong structural framework from a natural material with a clean and flexible construction procedure.

4.1 (previous page) Composite wood provokes a distinct dialogue between craftsmanship and modernity.
4.2, 4.5 In the production of glued laminated timber, wooden strips are bonded under heat and pressure to form a single member.
4.3, 4.4 The composition of a glued laminated beam is more technologically advanced than its appearance suggests: strong wood can be placed at the top and bottom of a section, like flanges in a steel beam.

4.5

AARAU MARKET HALL

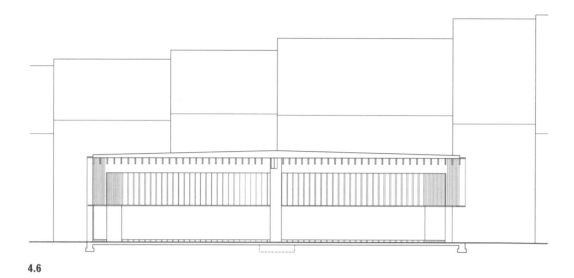

4.6

The market hall in Aarau by Quintus Miller and Paola Maranta is a building as a spatial lamination. The material concept of glued laminated timber unifies all features of the project into a common theme, relating the building to its cultural environment in multiple ways. The hall, which was completed in 2002, shelters vendors from the elements while providing a level ground for the market stalls. On the exterior, its level floor is used as a raised plaza for communal recreation. The open space of the plaza and the roofed space of the hall distill the traditional activities of a European town square into distinct architectural elements.

URBAN LAMINATION

The tall and narrow townhouses on the site create an urban lamination that resonates in the architecture of the hall and in the composite of its glulam members.

The building is glued together from regular slices of space and then trimmed into its skewed shape to fit into its urban environs. The resulting structure is conceptually form-less; its irregular geometry is determined by the medieval street pattern of the town. Both long facades of the hall are parallel to the adjacent street-walls, offset only by the typical alley width. A passage through the surrounding houses determines the front end. The hall is not an autonomous object that stands in the public space; it is the public space of the marketplace articulated as a built volume.

The tectonics of this building refer to the history of wood construction in a direct way. Its combination of major and minor members bears an enlarged resemblance to the canon of a traditional timber structure. In its unpretentious appearance the hall references the

4.7

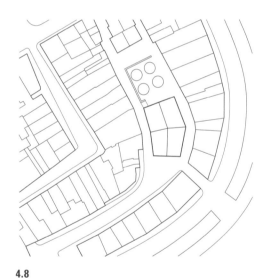

4.8

4.6, 4.7 The narrow townhouses on the site create an urban lamination that resonates in the glulam members of the market hall. They vary in size and function, but maintain a consistent shape and appearance throughout.

4.8 The architecture of the hall is glued together from regular slices of space and trimmed into a skewed shape to fit into its urban environs.

effortless beauty of agricultural buildings like barns or mills. Yet, the wood is used in the form of an advanced composite. A contemporary expression of engineered design and industrial production complements the rural connotations.

Glued laminated timber is used in several ways to compose this modern market hall: The beams demonstrate how the material can be used as a horizontal bending member; the posts show how it performs vertically in compression; other members show how it can be trimmed to a specific section-profile. As a result, the space of the market is a giant arrangement of linear wood pieces that vary in size and function, but maintain a consistent shape and appearance throughout.

The assembly of a crooked building framework from straight members requires a distinguished structural logic. A large post stands at the midpoint of

each of the four perimeter walls, breaking the bent shape down into straight segments. These four posts transition at the top into girders that intersect over a large central column. Assembled from four glulam members into a square tube, this column establishes a structural and spatial nucleus for the building. It provokes the archaic symbolism of a single pole or tree standing in the center of a marketplace. Indeed the structural system functions like a tree, where the major girders branch out to four sides from the central support. A longitudinal girder that carries all secondary beams functions as the structural spine of the roof. Across the short axis of the building another girder is introduced to stiffen the column laterally. It is less deep than the longitudinal girder since it does not support any roof beams.

As typical for glulam, the central girder is composed of horizontal laminations that react to vertical loads. Each layer is two inches thick, adding up to a total depth of three feet. Two wide members of laminated timber sheathing are screwed to the top of the girder. They connect it to the adjacent plywood ceiling panels for a strong bracing against lateral deformation. The ceiling panels are notched into the top of the girder and rigidly bonded to all secondary beams, to stiffen the entire structure. The whole roof becomes a composite wood product that is bonded together from various laminated components. Spacer pieces on top of the plywood ceiling generate the pitched angle of the roof surface. This allows the beams to remain straight. Placing the glulam beams at an angle rising towards the

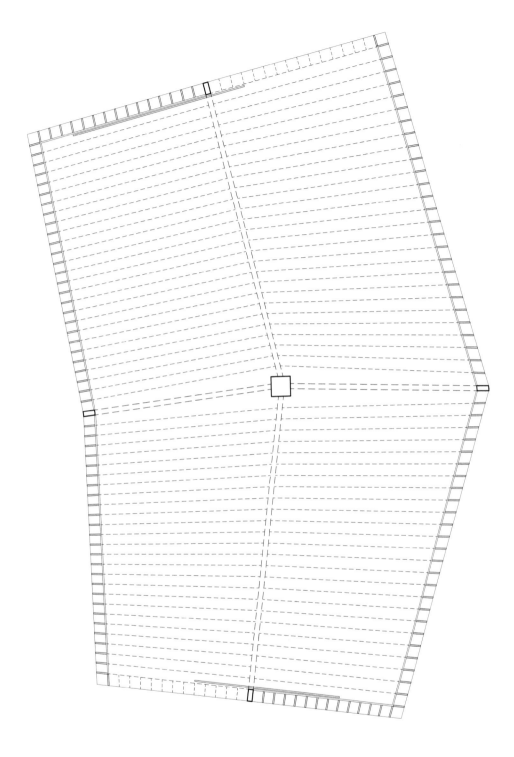

4.9, 4.10 The structure functions like a tree, where
four girders branch out from a central column to large
posts at the midpoint of each perimeter wall.

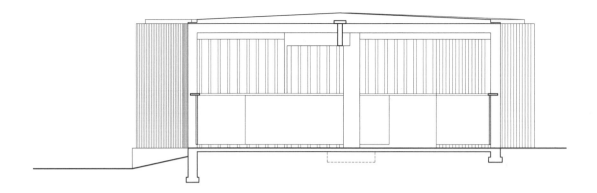

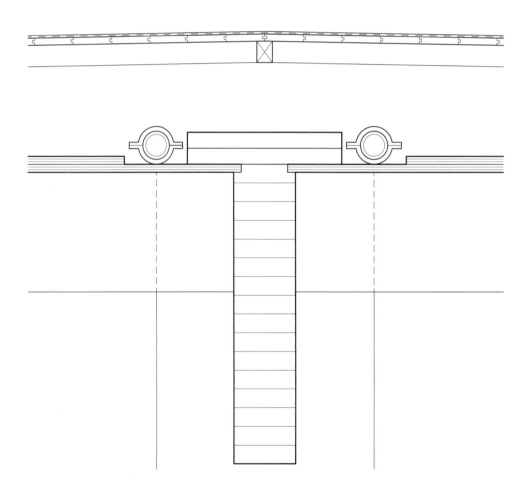

Composite Wood / Case Study

4.13

center would lose the abstract spatial gesture of the hall. Aside from the desirable appearance, this solution was also more economical: Rather than producing many different trapezoidal shapes, one beam is dimensioned for the maximum span and cut to the required length for its specific location.

4.11 A longitudinal girder functions as the structural spine of the roof. Across the short axis of the building another girder stiffens the column laterally.
4.12, 4.13 The longitudinal girder supports the secondary beams and is rigidly bonded to the plywood ceiling. This makes the entire roof a composite wood product that is bonded together from various laminated components.
4.14 (overleaf) On the long walls, the perimeter posts bend at the roof line to become a beam. When seen frontally, the narrow side of the posts is visible, giving the wall a transparent appearance.

TRANSPARENT AND OPAQUE

The roof could have been built with stronger beams spanning directly from wall to wall across the entire space, without the longitudinal girder and its column support in the center of the hall. But this larger span would have required a heavier structure at the perimeter of the hall. The main girder in the center eliminates the necessity of a thick edge beam at the roofline, allowing for a lighter, almost ephemeral facade. In the lower zone of the wall, panels attach to the inside of the structural framework, creating an enclosed interior. The upper part of the perimeter wall is open, consisting only of slim glulam posts. From across the alley the narrow sides of the posts is visible, and the boundary of the

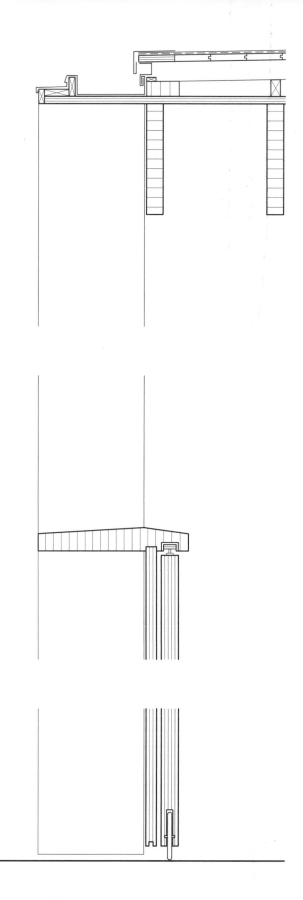

Composite Wood / Case Study

wall is transparent. The neighbors maintain their view of the public square from the upper floors of the adjacent buildings. The only difference is that they are now looking onto it through a screen of slender posts. When seen from the outside in a steep angle, the wide side of the posts is visible and the façade becomes opaque throughout. This makes the closed lower portion and the open upper portion appear as a single volume.

A horizontal rail separates the closed lower zone of the wall from the open upper part. It provides lateral stability for the slender posts, functioning like

4.15 A horizontal rail separates the closed lower zone of the wall from the open upper part. It provides lateral stability for the slender posts. The roof ends at the inner sides of the posts, creating a sharp top edge for the wall.

4.16 On the short walls, the perimeter posts do not bend at the roofline to become a beam. They form a screen and support the rail that holds the sliding door.

a beam turned on its side. The resistance to horizontal forces is demonstrated by the vertical orientation of the lamination layers. A trimmed profile gives the rail a slender appearance on the interior and exterior elevations. Its slanted surfaces shed rainwater, and define a sharp boundary between the interior and the exterior of the hall. The crest of this division occurs along the inside edge of the posts, conceptually attaching the posts to the outside of the inhabitable space.

The roofing layers stop at the inner side of the perimeter posts; they do not extend to the outside edge of the building. This creates the sharp edges of the wall along the roofline. The roof seems to have no thickness when viewed from street level. A copper gutter on top of the posts collects the rainwater from the roof. This metal element is integral to the surface treatment of the entire building: all the timber members are finished with copper

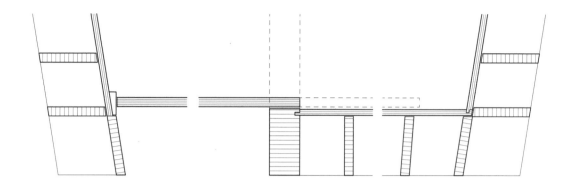

sealant for weatherproofing, which gives the structure its colorful glow; the gutter is a condensed version of this coating.

The assembly of members oscillates between the logic of a linear structure and a square structure. The overall plan is a skewed rectangle with roof beams running perpendicular to its long sides. In this directional system, the beams are supported by the longitudinal girder and by the thin posts in the long walls. Yet these posts run around all sides of the building as if the roof was a square structure spanning in both directions. The dialog between directional and non-directional aspects is expressed in the variations of the perimeter posts. On the short end walls, they do not bend at the roofline to become a beam. Instead they form a screen and support the rail that holds the sliding door. Although similar in size, shape, and spacing, a thinner section profile expresses the different structural function of these posts.

The beams and their posts run in parallel along the main axis of the building. This makes them meet the perimeter in oblique angles. The short ends of the posts are trimmed accordingly to create a flush wall edge.

LAMINATED SPACE

The outer faces of the posts are cut in angles to fit the geometry of the structure, and the structure as a whole is cut in these same angles to fit the urban site. The entire building is one composite wood product made from individual composite wood materials. Its interior volume is a series of vertical spatial layers that are sliced in one plane along the perimeter. Within the hall, layers of adhesive connect the individual laminations of the glulam members, as the glulam members glue together the slices of its space. The regular vertical organization of the architecture resembles the internal organization of the material, and the irregular shape of the building demonstrates the effortless customization of the material by cutting it into a specific shape.

4.17 Although similar in size, shape, and spacing, a thinner section profile expresses the different structural function of the posts in the front façade.
4.18 When seen in a steep angle, the wide side of the posts becomes visible and the façade is opaque throughout.

PLYWOOD

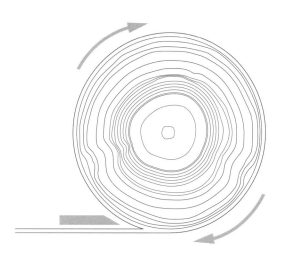 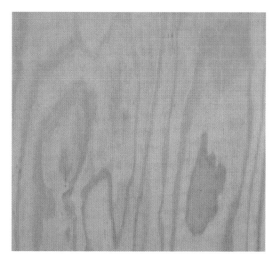

Like glulam, *plywood* is composed of individual wood pieces bonded together with adhesives. Already in ancient Egypt builders had responded to shortages of quality lumber by attaching thin wooden sheets to a single plank. In the nineteenth century the Swedish engineer and industrialist Immanuel Nobel realized that several thinner layers of wood bonded together create a stronger wood surface than a single thick layer. He invented the *rotary veneer* cut that shaves a *ply* of a desired thickness from a log by rolling it along a knife. This ply is cut into a rectangular sheet called a *veneer*. Pairs of veneers are glued to the two sides of a center ply, the *core*. In their arrangement, the grain of each ply runs perpendicular to the plies above and below. This neutralizes the directionality of the wood and makes the composite equally strong in either orientation. The opposite grains also neutralize the tendency each veneer has to bend when it absorbs moisture. Plywood is always composed of an odd number of plies; the grain of the outmost plies called *face* and *back* run in the same direction accordingly. The type of adhesive used to assemble the plies determines if the plywood can be used on the exterior where it has to resist moisture, or if it is only suitable for interior application.

Plywood transforms organic material into a standard rectangular form, typically eight feet by four feet. In Europe the similar metric format is two hundred forty by one hundred twenty centimeters. The rectangular proportions are determined by the use of the product in construction: eight feet is a typical floor-height, four feet is a typical door width. When used as flooring,

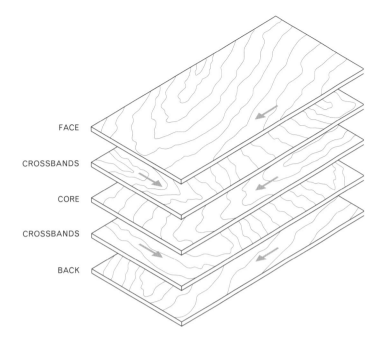

FACE

CROSSBANDS

CORE

CROSSBANDS

BACK

the short dimension of the plywood can transfer loads to linear supports. Eight feet by four feet is also the largest sheet size that can be handled well by an individual carpenter on site.

As a lamination of wooden veneers, plywood maintains the layered characteristic of natural timber. The concentric year rings of a tree are unwound with the rotational veneer cut. When reassembled as a rectangular sheet, the plies surrounding the core again resemble the growth of a tree. Manufacturing plywood is thus a reversal of a tree's growth that is reassembled in a rectilinear logic. The resulting product has a natural consistency and an artificial form.

4.19 A rotary veneer cut shaves plies of a desired thickness from a log by rolling it along a knife.
4.20 The ply is cut into a rectangular sheet called a veneer.
4.21 Pairs of veneers are glued to the two sides of a center ply. The grain of each ply runs perpendicular to the plies above and below, neutralizing the directionality of the natural wood fiber.
4.22 Plywood is typically composed of five, seven or nine plies.

GÖTZ COLLECTION

The Götz Collection in Munich designed by Jacques Herzog and Pierre de Meuron with Helmut Federle takes the character of plywood as a guiding logic of its design. This small museum of contemporary art was commissioned by a private collector, and was completed in 1992. The use of plywood informs the architectural idea of the exhibition building from initial design idea through execution. Birch wood, commonly used in making plywood, is the dominant material of this project. Three different incarnations of this material are manifest:

first as the birch trees that stand in front of the museum, second as a composite wood product, and finally as a distinct architectural form.

ART CRATE

The museum enlarges the form of an art crate to the scale of a building. Used for the storage and shipping of artwork, these containers are usually constructed from plywood sheets held together by a wooden framework. They are customized for specific works of art, referring to the size and proportion of the objects they envelop, and carry the semblance of their precious content. Art crates are a ubiquitous feature of art installations and exhibitions. In the Götz Collection, the art crate becomes the museum itself. The structural grid of this architectural container is based on the industrial sizes of plywood. The standard dimensions of this material determine the proportions of the building. Multiple sheets of plywood assemble a larger rectangular volume that is held together by a glulam frame. The entirety of this timber structure sits atop two concrete blocks, which are slightly set in from the short ends of the building; they are positioned independently of the structural grid of the plywood box, underlining its tectonic autonomy.

The art crate has an immediate quality, as if it had just arrived at the site and was in the process of being unpacked. Its lid is lifted up as a floating roof, creating a clerestory at the top of the building. Here, a row of frosted glass sheets replaces the plywood panels. The same kind of glass closes the gaps between the concrete blocks under the plywood box. During mild daylight, glass and wood have a similar matte appearance, unifying the building as a single, solid volume. At night, when the galleries are illuminated from the inside, the façade is sharply divided between the translucent glass and the opaque plywood.

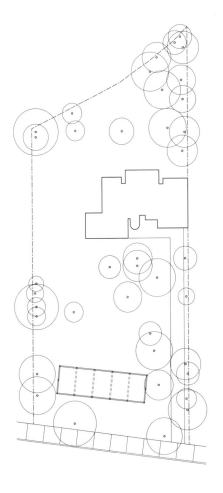

4.23 The Götz Collection enlarges the form of a plywood crate used for the storing and shipping of art to the scale of a building.

4.24 Art crates are constructed from plywood sheets held together by a wooden framework. They refer to the size and proportions of the artwork they envelope and carry the semblance of their precious content.

4.25 Situated in a garden between a house and the street, the museum building can be used as a private pavilion and as a public exhibition space.

Götz Collection

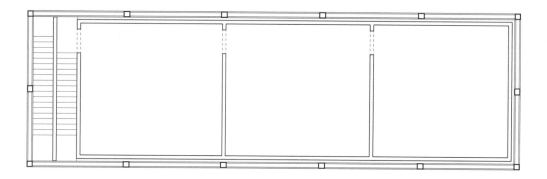

202 Composite Wood / Case Study

4.27

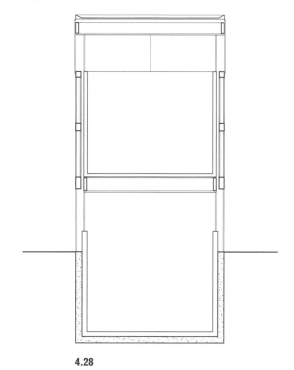

4.28

The Götz Collection is located in the garden of a private residence. It shares the orientation of the house, but is slightly rotated to be parallel to the street. As a result, it is perceived from the sidewalk as an arrangement of plywood and glass planes, rather than as

4.26 The upper and lower galleries are identical in size and proportion, yet distinct in construction and relation to the ground: in the lower level (bottom) a concrete trough is buried in the earth. The only inhabitable spaces on ground level (center) are two concrete tubes that support the timber structure of the upper level (top).

4.27, 4.28 What appears from the outside as a glazed ground floor is actually a clerestory for the gallery below.

4.29 (overleaf) Because of their similar shape, the dialogue of different materials in the two parts of the building becomes tangible. When the galleries are illuminated from the inside, the façade is sharply divided between the translucent glass and the opaque plywood.

a three-dimensional object. The location between house and the street allows the building to be used as a private pavilion that can be approached from the residence, and as a public exhibition space that can be accessed directly from the street. Consequently, the entrance vestibule has two open sides facing towards the house and street respectively.

CONCEPT ART – CONCEPT ARCHITECTURE

The different levels of the building express a specific set of relationships between form and materiality. What seems to be an open ground floor below the lifted plywood box, is actually another clerestory for a second underground gallery. The only inhabitable spaces at ground level are inside the two concrete tubes that support the timber structure above.

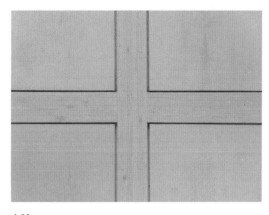

4.30

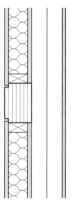

4.31

Both of them span across a concrete trough embedded in the ground that houses an additional sequence of gallery rooms; these spaces maintain the same proportions as the exhibition spaces in the lifted plywood box. Both galleries are lined with sheetrock, enveloping the artwork like the styrofoam or paper padding surrounding an object in an art crate. Throughout the building this inner sheetrock wall layer interacts with the outer façade in different degrees of independence. In the upper level it is connected to the glulam structure with metal ties. On the lower level it attaches directly to the concrete surface. The sheetrock rises above the edge of the concrete trough, forming a parapet within the glazing. The top of this parapet defines the height of the clerestory for the lower gallery, making it identical to the clerestory of the upper gallery.

Although identical in size and proportion, the two galleries are distinct in construction and relation to the ground. The different materials used for the creation of the same spatial configuration emphasize the contrast of surfaces in spite of the consistency of interior space. The lightness of the plywood structure is demonstrated in how it is lifted above the ground; the strength and heaviness of the concrete is expressed in the way it is embedded into the ground respectively. This sculptural elaboration of form and structure embodies the qualities of each material in the same prismatic shape. Because of their identical proportion, the dialogue of materials and building methods in the two parts of the building becomes tangible.

THE GALLERY AND ITS OTHER

The upper and lower galleries with the same volume but contrasting expression are conceptually *others* of each other. The concrete trough constitutes the identity of the plywood crate and vice versa by being different and at the same time identical. All characteristics of the project are derived from this concept, avoiding any additional architectural features. The reduction of aesthetic design decisions allows the architects to arrive at a building that is formally detached from the hand of its creator

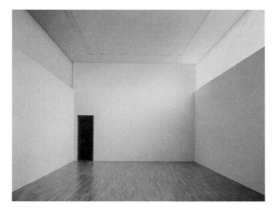

4.32

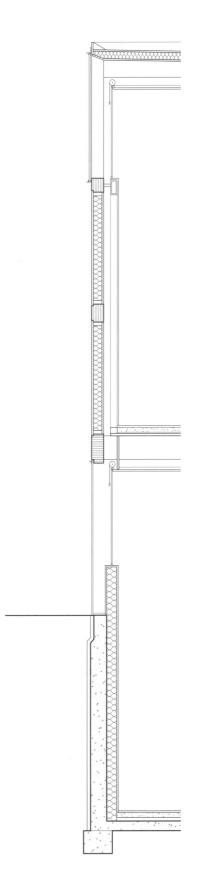

and solely represents the essence of its idea. Herzog, de Meuron and Federle restrict the number of formal decisions to a minimum. They deny any personal expression in the appearance of the building, in favor of one artistic concept; every feature of the project becomes entirely abstract.

The plywood sheet is used as it is. Through the idea of the art crate it becomes possible to use the industrial material as the generating force of a museum design. Refining construction details to a level of abstraction that caters to the metaphor of the enlarged plywood box, while still providing an inhabitable building is a major challenge. The clarity of the idea demands a sense

4.30, 4.31 Because plywood is strong in shear, it can contribute to the stiffness of the wall. This new structural capacity of cladding is expressed in the minimized joints of the glulam framing.
4.32, 4.33 The upper and lower galleries with the same volume but contrasting expression are conceptually others of each other. They feature the identical display walls and clerestory windows although one is buried in a concrete trough and the other is lifted in a timber box.

Composite Wood / Case Study

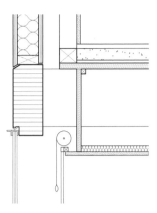

of construction that is modern; all elements of the building express their role within the structural assembly. At the same time, the construction also represents an artistic concept that is disconnected from the scale of its human inhabitation.

CLADDING AND FRAME

The structural system of the Götz Collection building changes the traditional relation between a timber frame and its cladding. Because plywood is strong in shear, it makes a cladding that can contribute substantially to the structural integrity of the building. It is more than a panel filling a frame; it also stiffens the entire structure as a *diaphragm* against lateral deformation. Consequently, the joints between the glulam members that assemble the frame do not have to create the rigidity of the structure. This is a revolutionary change in timber construction. Where the structural emphasis had once been upon the connection of the linear frame members, it is now focused on the connection between frame and cladding.

Plywood panels are connected to the glulam framing by a nailer piece fastened to the top and bottom of the beam. This nailer piece is narrower than the beam by exactly twice the thickness of the plywood, allowing for the precise alignment of the panels with the faces of the beam. In elevation this results in two perfectly aligned surfaces separated by a precise shadow reveal. The outer composite wood part of the wall consists of two plywood sheets fixed to both sides of the nailer piece to create the necessary rigidity; the thermal insulation of the wall is sandwiched between them.

The new role of the plywood in the structural system is expressed by the fact that it determines the construction grid of the building. While the dimensions of

4.34, 4.35, 4.36 The large glulam girders were brought to the site as a single piece with the connecting hardware for the floor joists already installed. The ends of the girder are cut in a 45 degree angle in order to connect at the corners with a miter joint that does not reveal the cross section of the member. A crane lifted the girder directly from the truck into their final position on the concrete supports.

4.37, 4.38 One large girder collects the entire load from the timber structure of the upper gallery and transfers it to the concrete supports below. While the dimensions of the plywood sheets remain consistent, the glulam members vary depending on their loading.

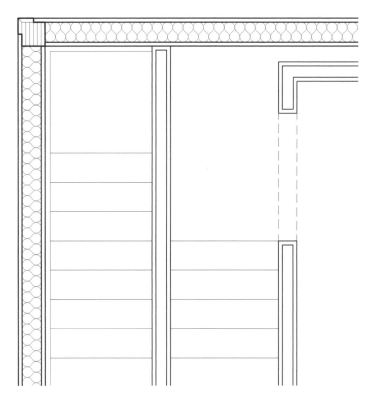

the plywood sheets remain consistent throughout, the glulam members vary depending on their loading. Under the clerestory at the top of the upper gallery wall a glulam beam consists of vertical layers; its structural function is to resist lateral loads. At the floor level of the upper gallery a large glulam girder collects the entire load from the timber structure and transfers it to the concrete supports below; this girder consists of horizontal laminations for vertical loading.

While the expression of the different forces at work in the material ties into a minimalist conception of the structure, additional features that facilitate the use of the art crate as a museum building are more difficult to integrate. The door openings between the individual

exhibition rooms demonstrate this conceptual limit. They seem out of place because they are neither derived from the art-crate metaphor, nor from the material assembly. The human body and the programmatic use of the space solely determine the door's size, proportion and location. These doors indicate the difficulty of inhabiting a built metaphor. If the architectural logic of a building is solely based on concepts about materials and their relation to the ground, the introduction of elements that facilitate inhabitation questions the integrity of the project.

The staircase that connects between the different levels is also introduced to facilitate the inhabitation of the art-crate; but it is integrated more successfully in the overall concept as a

tectonically independent element. It is neither a part of the concrete trough or the timber box; it also maintains an autonomy from the inner and outer structural layers of the wall in each level. The white lining stops and turns inward to stand by itself as an independent wall where the exhibition rooms end and the staircase begins. From the staircase on the way to the lower gallery, the exposed underside of the wooden container is visible free of any supports; the thickness of the sheetrock lining inside the frosted glass becomes visible. On the upper level the independence of the plywood shell is

4.39, 4.40, 4.41 (overleaf) The staircase that connects between the different levels is neither a part of the concrete trough or the timber box. It is the only moment in which the glulam and plywood assembly of the art crate can be experienced from inside the box.

prominently displayed. This is the only spatial moment in the gallery in which the assembly of glulam frame and plywood panels can be experienced from inside the box.

COMPOSITE VERSUS JOINT

The architectural concepts of both the market hall in Aarau and the Götz Collection in Munich emerge from the characteristics of the artificial timber products used for their construction. Both buildings make tectonic gestures explicitly about the structural capacities of composite wood products, referring to the logic of the composite rather than the logic of the joint. In glulam the large members themselves can be customized to a desired shape or function; there is no need for complex timber joints.

Composite Wood / Case Study

The market hall represents this total customization of any kind of wood member. It celebrates the irregular shape of its architecture in the trimmed shape of its components. On the opposite end of the spectrum, the Götz Collection celebrates the regularity and standardized dimensions of plywood. It creates an architecture based on the uniformity of an industrial product that is assembled to form a pristine geometric volume.

Both the extreme customization and extreme regularity are made possible by the precise engineering of wood products. In these projects assembly has already begun with the gluing procedure in the factory, and construction is only a further step in the sequence of creating the wood composite. This new combination of industrial prefabrication and traditional carpentry results in a strictly contemporary expression of wooden architecture. In the new timber construction, members of all scales participate in the structural force-flow; there is no longer a separation between members that are carrying and members that are being carried. But the natural appearance of the engineered material converses with historic predecessors in wooden buildings. Although composite wood transcends its natural form in becoming an industrial product, glulam and plywood still embody the memory of traditional architectures that have been constructed from timber throughout the ages.

ACKNOWLEDGEMENTS

Like a good building, this book is the result of great teamwork. When I started to teach seminars on the relation between construction innovations and architectural form at The Cooper Union in 2009, a group of students encouraged me to write a book based on my lectures. As alumni, they became the co-editors of this publication: Lisa Larson-Walker, Sean Gaffney and Will Shapiro.

As a group, we developed the concept of the book in countless conversations not only about architecture and engineering, but also about the arts and sciences, philosophy, education and finally bookmaking. We were enthusiastic about the idea of creating an "anti-textbook" on construction technology, meant to inspire rather than instruct.

All of us are grateful for the patience of our friends and families that helped us realize a book of this size over the course of five years without any institutional support. I am deeply thankful to my wife Lydia Xynogala, my parents Hasso and Alice Windeck and my sister Nora Windeck for all their tremendous support throughout these years.

Many ideas in *Construction Matters* are inspired by a continuous exchange with my friend and colleague Daniel Schütz. His criticism and support fundamentally impacted the form and content of this book from the initial concept phase through its final stages.

Other friends and former students made tremendous contributions to the visual material in the book: Dennis Gilstad and Ivan Himanen traveled far into Japan and Finland respectively to take photos of case study buildings. Frank Seehausen researched archives in Berlin to gather historic photos and Joseph Vidich of Kin&Company manufactured special welding samples for the steel chapter. Many colleagues and institutions generously provided us with existing photos. Rolf Weisse made his distinguished construction photos of the New National Gallery available to us and Werner Blaser gave permission to use his photograph of the Farnsworth House. The Götz Collection provided us with pictures of their building in Munich and the Acropolis Restoration Service provided pictures of the Parthenon and Erechteion in Athens.

Many text passages in the book have been peer reviewed by distinguished scholars and professionals such as Roger Conover, Robert Hullot-Kentor, Holger Kleine and Toyo Ito. Their comments gave new insights and sharpened the content of the book from various angles.

Along the way many other people supported our efforts in spontaneous ways that helped us keep going. We owe our special thanks to Alexandra Alexa, Roger Duffy, Matthew Harvey, Mindy Lang, Mary Lynch, Monica Shapiro and Tamar Zinguer. The inspiring collaboration with the artist Leah Beeferman allowed us to condense the various visual and written components of the book into a comprehensive graphic design.

Finally we thank all the backers of our fundraising campaign for their enthusiasm and support; we could not have produced this publication without your help! We are especially grateful for the generous contributions of Page Ashley, Eunil Cho, Tse-En Fan, Terence Gaffney, Thomas Mog and Robert Shapiro. Without their support it would not have been possible to publish *Construction Matters*.

New York, September 2014

INDEX

Construction Matters

Construction Matters

BIBLIOGRAPHY

INTRODUCTION

Konstruktion und Form im Bauen
(Construction and Form in Building)
Friedrich Hess (Editor), Verlag Julius
Hoffman, Stuttgart 1949.

Le Corbusier Oeuvre complete
(Complete works)
Willy Boesiger (Editor), Editions Girs-
berger, Zurich, 1929–70.

Bauen Wohnen Denken (Building
Dwelling Thinking)
Martin Heidegger, in: Vorträge und
Aufsätze, Verlag Günther Neske,
Tübingen, 1954; English version translated by Albert Hofstadter in: *Poetry,
Language and Thought*, Harper Colo-
phon Books, New York, 1971.
In his lecture held at the Darmstädter
Gespräche (Darmstadt Talks) in 1951,
Heidegger discusses the origin of the
word "building" as related to that of
"being" and "farming." He concludes
that building should be understood as
a continuous process. It is not a singular
act preceding the inhabitation of architecture, but rather an integral part of
dwelling, and of existence as such.

Die vier Elemente der Baukunst
(The four elements of architecture)
Gottfried Semper, in: *Wissenschaft, Indus-
trie und Kunst*, Verlag Friedrich Vieweg
und Sohn, Braunschweig, 1852; English
version translated by Harry F. Mallgrave
in *The Four Elements of Architecture
and Other Writings*, Cambridge University Press, Cambridge, 1989.
Gottfried Semper analyzes the origins
of architectural form through primal
aspects of inhabitation. He divides architecture into four distinct elements that
are related to a traditional craft and
construction material of ancient "barbar-
ians:" the hearth (fire and ceramics), the
roof (carpentry), the enclosure (weaving)
and the mound (stone masonry).

*Studies in Tectonic Culture: The Poet-
ics of Construction in Nineteenth and
Twentieth Century Architecture*
Kenneth Frampton, John Cava (editor),
The MIT Press, Cambridge, 1995.

*The Details of Modern Architecture,
Volumes 1 & 2*
Edward Ford, The MIT Press, Cam-
bridge, 1990, 1996.

BRICK MASONRY

*Mies Van Der Rohe: Die Kunst der
Struktur* (The Art of Structure)
Werner Blaser, Artemis Verlag, Zurich,
1965; English version translated by
D. Q. Stephenson, Frederick Praeger
Publishers, New York, 1965.
The architect Werner Blaser worked for
several years with Ludwig Mies van der
Rohe. Here he discusses Mies' buildings
and analyzes his theories of structure
with photographs and drawings. The

combination of visual material and text give powerful insight into Mies' way of thinking about the relation between architectural form and structural design.

Das Kunstwerk im Zeitalter seiner technologischen Reproduzierbarkeit (The Work of Art in the Age of Mechanical Reproduction)
Walter Benjamin, in: *Gesammelte Schriften*, Rolf Tiedemann, Hermann Schweppenhaeuser (Editors), Suhrkamp, Frankfurt am Main, 1955; French version translated by Pierre Klossowski in: *Zeitschrift für Sozialforschung #5*, Félix Alcan, Paris, 1936; English version translated by Harry Zohn in: *Illuminations*, Hannah Arendt (Editor), Knopf Doubleday Publishing, New York,1968.

Alvar Aalto
Karl Fleig, Frederick Praeger Publishers, New York, 1975.

THIN SHELL CONCRETE

Candela the Shell Builder
Colin Faber, Reinhold Publishing, New York, 1963.
This book on the work of the engineer and contractor Felix Candela contains his essay "The Hyperbolic Paraboloid." In it, Candela talks about the mathematical properties of the double curvatures he uses in the structural design of his thin concrete shell projects. A detailed description of his construction process illuminates his innovative and economic approach to creating large span roofs from reinforced concrete.

Formprobleme der Gothik
(Form Problems of the Gothic)
Wilhelm Worringer, M. Pieper & Co., Munich, 1911.
In this book Worringer, an art historian, celebrates the gothic impulse to create art and architecture that is "stylized." This tendency reveals the psychological need to represent objects in a spiritual manner, as opposed to the realistic representation of ancient Greek and Roman art that demonstrated a confidence in the material world. Gothic architecture displays an insecurity with materialism, putting trust in spirituality through the abstract representation of structural form and ornamentation. It attempts to aesthetically overcome the physical presence of its structure to arrive at a spatial experience that transcends the material world.

Evolutionary Structural Optimization
Mike Xie and Steven Grant; Springer Verlag, Vienna, 1997.

STEEL FRAMING

Metal Shaping Processes
Vukota Boljanovic, self-published in New York, 2009.

Tragwerke in der Konstruktiven Architektur (Structural Systems in Constructive Architecture)
Kurt Ackermann, Deutsche Verlags-Anstalt, Stuttgart, 1988.

Der Stahlbau (The Steel Building)
Zeitschrift für Stahl-, Verbund- und Leichtmetallkonstruktionen (Journal for Steel-, Composite-, and Lightweight

Metal Construction), Ernst und Sohn, Berlin, 4/1968.

This journal contains the articles "Das Stahldach der Neuen Nationalgalerie in Berlin" (The Steel Roof of the New National Gallery in Berlin) by H. Oeter, and H. Sontag, and "Statische Untersuchungen für die Dachkonstruktion der Neuen Nationalgalerie in Berlin" (Static Analysis for the Roof Construction of the New National Gallery in Berlin) by K. Roik. These articles give detailed account of the static design, steel types and welding methods used for the fabrication of the New National Gallery roof components. It also describes the assembly procedure and hydraulic hoisting of the roof on site.

Mies van der Rohe Vision und Realität. Von der Concert Hall zur Neuen Nationalgalerie (Mies van der Rohe Vision and Reality. From the Concert Hall to the New National Gallery)
Rolf D. Weisse, J. Strauss Verlag, Potsdam, 2001.
The architect Rolf D. Weisse worked from 1964 to 1968 in the Chicago office of Mies van der Rohe on the design and construction of the New National Gallery. His book offers a personal insight into Mies' special working methods at his atelier and at the IIT Chicago.

STRUCTUREmag
Structure Magazine, publication of the National Council of Structural Engineers Associations, Chicago, 2/2007
This issue of the journal includes the article "Computer Technology in the Practice of Structural Engineering"
by Jim De Stefano. The author traces the history of computation in the engineering of buildings from the early structural analysis programs such as STRUDL that were developed at the Massachusetts Institute of Technology in the 1960s to today's Building Information Models.

The Architects of the Parthenon
Rhys Carpenter, Pelican Books:
The Architect and Society, Penguin Books Ltd., Baltimore, 1970.
Carpenter illuminates the architectural features and aesthetic refinements of the Parthenon through a compelling description of the building's construction history. The book discusses how the canonical rules of proportion from marble sculpture impacted architectural design, and how the "archi-tekton" of the Greek temple worked as a chief craftsman with the marble cutters on site.

De Architectura libri decem (Ten Books on Architecture)
Marcus Vitruvius Pollio, 1st Century BC, Rome; English version translated by Morris Hicky Morgan, Harvard University Press, Cambridge, 1914.

Machinery's Encyclopedia
Erik Oberg and Franklin Jones, The Industrial Press, New York, 1917.

Finite Element Modeling for Stress Analysis
Robert Cook, Wiley & Sons, New York, 1995.

GA Detail
Global Architecture, A.D.A. EDITA, Tokyo
1976: Mies van der Rohe, Farnsworth House, Plano, Illinois, 1945-50, text by Dirk Lohan.
2001: Toyo Ito & Associates, Sendai Mediateque, Miyagi, Japan, 1995-2000, Texts by Toyo Ito and Mutsuro Sasaki, translated from Japanese by Hiroshi Watanabe
The Imperial Grand Shrines at Ise in Japan are completely taken apart and rebuilt every twenty years. This ritual of Shinto Buddhism has led to the preservation of ancient Japanese architectural forms: every single member and joint of the shrine's timber structure is carefully documented in exhaustive catalogues for its precise reconstruction. The *GA Detail* series adopts this concept for the documentation of distinguished pieces of 20th century architecture, elevating them to temples of architectural construction.

This book tells the history of the Nobel family and discusses the inventions they contributed to the foundations of industrialization. Immanuel Nobel (1801-1872) invented the rotary veneer cut that shaves a wooden log into plies that can be glued together to create plywood sheets.

COMPOSITE WOOD

Laminated Timber Construction
C. Müller, Birkhäuser, Basel, 2000.

Timber Construction Manual
Thomas Herzog, Julius Natterer, Roland Schweitzer, Michael Volz, Wolfgang Winter, Birithäuser, Basel, 2004.

The Russian Rockefellers: The Saga of the Nobel Family and the Russian Oil Industry
Robert W. Tolf, Hoover Publications, Stanford, 1976.

IMAGE CREDITS

Every reasonable attempt has been made to identify authors and owners of copyrights. We apologize for any errors or omissions that may have occurred. Please contact us should you have any corrections.

COVER AND TITLE

0.1 Construction of steel roof, New National Gallery, Ludwig Mies van der Rohe, 1969, photo by Architect Rolf D. Weisse, Berlin, courtesy Rolf D. Weisse.

0.2 Farnsworth House, Ludwig Mies van der Rohe, 1949-50, photo by Construction Matters.

0.3 Welding samples by Joseph Vidich of Kin & Company, photo by Construction Matters.

INTRODUCTION

0.4 Monadnock Block, Chicago, 1898, Burnham and Root, photo by Construction Matters.

0.5 Parking garage, Chicago, ca 1970's, Architect unknown, photo by Construction Matters.

0.6 Shibaura Building, Tokyo, 2011, SANAA, photo by Dennis Gilstad, courtesy Dennis Gilstad.

0.7 Gamble House, Pasadena, 1908, Gamble & Gamble, photo by Amanny Ahmad, courtesy Amanny Ahmad.

BRICK MASONRY

1.0, 1.6, 1.8 Plan, plan details, Brick Country House project, 1924, Ludwig Mies van der Rohe, drawings by Construction Matters.

1.1, 1.2 Detail of Flemish brick bond, drawing and photo by Construction Matters.

1.3 Extruded wire cut brick production. Photomontage by Construction Matters.

1.4, 1.5 Perspective and plan, Brick Country House project, 1924, Ludwig Mies van der Rohe, photo of drawing by architect, courtesy Städtische Kunsthalle Mannheim.

1.7 Aerial view, Dominion Center, Toronto, 1967, Ludwig Mies van der Rohe, photo by Ron Vickers, TD Financial Group Archives.

1.9 Axonometric details of brick walls, Ludwig Mies van der Rohe, drawing by Construction Matters.

1.10, 1.11, 1.12, 1.13 Elevation, plan and construction details, House with Three Courtyards project, 1934, Ludwig Mies van der Rohe, drawings by Construction Matters.

1.14 Reinforced masonry construction, photomontage by Construction Matters.

1.15 Axonometric details of reinforced masonry construction, drawings by Construction Matters.

1.16 Frontal view, Monument to Rosa Luxemburg and Karl Liebknecht, 1926, Ludwig Mies van der Rohe, photo by unknown, MoMA, ARS, Art Resources.

1.17, 1.18 Elevation sketch and site plan drawing, Monument to Karl Liebknecht and Rosa Luxemburg, 1926, Ludwig Mies van der Rohe, MoMA, ARS, Art Resources.

1.19 Corner view, view, Monument to Rosa Luxemburg and Karl Liebknecht, 1926, Ludwig Mies van der Rohe, photo by Arthur Koster, MoMA, ARS, Art Resources.

1.20, 1.21 Elevations, Monument to Karl Liebknecht and Rosa Luxemburg, 1926, Ludwig Mies van der Rohe, drawings by Construction Matters.

1.22, 1.27, 1.29 Site plan, elevation, courtyard plan, House in Muuratsalo, 1953, Alvar Aalto, drawings by Construction Matters.

1.23, 1.24, 1.25, 1.26, 1.28, 1.30 Exterior and courtyard views, House in Muuratsalo, 1953, Alvar Aalto, photos by Ivan Himanen, courtesy Ivan Himanen.

THIN SHELL CONCRETE

2.0 Roof plan detail, Meiso no Mori Municipal Funeral Hall, Toyo Ito and Associates, drawing by Construction Matters.

2.1 Roof detail, Trans World Flight Center, 1962, Eero Saarinen, photo by Construction Matters.

2.2, 2.3, 2.4 Compound surfaces, membrane stresses, warped shapes, diagram drawings by Construction Matters.

2.5, 2.17 Exterior view, interior view, Los Manantiales Restaurant, 1958, Joachin and Fernando Alvarez Ordonez with Felix Candela, unknown photographer, Félix and Dorothy Candela Archive, Princeton University.

2.6, 2.7, 2.8, 2.9, 2.11 Plan, elevation and section, site plan, groin vault sections, Umbrella footing detail Los Manantiales Restaurant, 1958, Joachin and Fernando Alvarez Ordonez with Felix Candela, drawings by Construction Matters.

2.10 Groin vault detail, Los Manantiales Restaurant, 1958, Joachin and Fernando Alvarez Ordonez with Felix Candela, unknown photographer, Félix and Dorothy Candela Archive, Princeton University.

2.12 Construction of umbrella footing, Los Manantiales Restaurant, 1958, Joachin and Fernando Alvarez Ordonez with Felix Candela, unknown photographer, Félix and Dorothy Candela Archive, Princeton University.

2.13, 2.14, 2.15, 2.16 Construction of roof formwork, reinforcing and concrete placement, Los Manantiales Restaurant, 1958, Joachin and Fernando Alvarez Ordonez with Felix Candela, unknown photographer, Félix and Dorothy Candela Archive, Princeton University.

2.18 Stress distribution in concrete roof, Meiso no Mori Municipal Funeral Hall, 2006, Toyo Ito and Associates, diagram drawing by Construction Matters.

2.19, 2.20, 2.21, 2.25 Elevation, site plan, cross section, ground floor plan, Meiso no Mori Municipal Funeral Hall, 2006, Toyo Ito and Associates, drawings by Construction Matters.

2.22, 2.23, 2.26, 2.42 Exterior and Interior views, Meiso no Mori Municipal Funeral Hall, 2006, Toyo Ito and Associates, photos by Dennis Gilstad, courtesy Dennis Gilstad.

2.24, 2.27, 2.35 Column and roof details, Meiso no Mori Municipal Funeral Hall, 2006, Toyo Ito and Associates, drawings by Construction Matters.

2.29, 2.37 Assembly column formwork, roof reinforcing plans, Meiso no Mori Municipal Funeral Hall, 2006, Toyo Ito and Associates, drawings by Construction Matters.

2.28, 2.30, 2.31, 2.32, 2.33, 2.34, 2.35, 2.38, 2.39, 2.40 Formwork assembly, reinforcing, concrete placement, formwork removal, Meiso no Mori Municipal Funeral Hall, 2006, Toyo Ito and Associates, photos by Toyo Ito and Associates, courtesy Toyo Ito and Associates.

2.41 Construction sequence overview, Meiso no Mori Municipal Funeral Hall, 2006, Toyo Ito and Associates, photos by Toyo Ito and Associates, courtesy Toyo Ito and Associates.

STEEL FRAMING

3.0 Column and glazing detail, Farnsworth House, Ludwig Mies van der Rohe, 1949-50, drawing by Construction Matters.

3.1, 3.9, 3.15, 3.17, 3.19, 3.37 Exterior views, interior views and details, Farnsworth House, Ludwig Mies van der Rohe, 1949-50, photos by Construction Matters.

3.2, 3.3 Rail steel profile, rolled steel profiles, drawings by Construction Matters.

3.4 Structural shape rolling mill, industry photo.

3.5, 3.6 Structural shape rolling, diagram drawings by Construction Matters.

3.7, 3.18, 3.20, 3.36, 3.49 Plan, elevation, section and details, Farnsworth House, Ludwig Mies van der Rohe, 1949, drawings by Construction Matters.

3.8, 3.14, 3.22, 3.51 Plan, ceiling plan, elevation and details, New National Gallery, Ludwig Mies van der Rohe, 1969, drawings by Construction Matters.

3.10 Structural steel profiles used in Farnsworth House, Ludwig Mies van der Rohe, 1949-50, drawing by David Ross, courtesy David Ross.

3.11 View across river, Farnsworth House, Ludwig Mies van der Rohe, 1949-50, photo by Werner Blaser, courtesy Werner Blaser.

3.12, 3.13, 3.16, 3.21, 3.35 Exterior and interior views, details, New National Gallery, Ludwig Mies van der Rohe, 1969, photos by Construction Matters.

3.23, 3.24 Plan of roof steel plates, axonometric of column and roof steel plates, New National Gallery, Ludwig Mies van der Rohe, 1969, drawings by Construction Matters.

3.25 Arc welding, photomontage by Construction Matters.

3.26, 3.27 Arc welding circuit, weld joint types, drawings by Construction Matters.

3.28, 3.29 Fillet welded T-joint and plug-welded lapped joint, welding samples by Joseph Vidich of Kin & Company, photos by Construction Matters.

3.30 Axonometric of steel frame, Farnsworth House, Ludwig Mies van der Rohe, 1949-50, drawing by Construction Matters.

3.21 Steel frame construction, Farnsworth House, Ludwig Mies van der Rohe, 1949-50, unknown photographer, Myron Goldsmith fonds, Collection Centre Canadien d'Architecture/Canadian Centre for Architecture, Montreal.

3.32, 3.33, 3.34 Elevation of column and column details, New National Gallery, Ludwig Mies van der Rohe, 1969, photo and drawing by Construction Matters.

3.38 Birds and Flowers of the Four Seasons, ink paintings on sliding wall panels in the Abbots Quarter of Daitoku-Ji, Kyoto. Attributed to Kanō Eitoko and his father Kanō Shōei. National Treasure of Japan, public domain.

3.39 Interior view of New National Gallery, Ludwig Mies van der Rohe, 1969, photo by Reinhard Friedrich, 1971, Staatliche Museen zu Berlin - Preußischer Kulturbesitz.